International Conference on Human Factors 2020

London, United Kingdom
19 – 20 February 2020

ISBN: 978-1-7138-0866-4

Printed from e-media with permission by:

Curran Associates, Inc.
57 Morehouse Lane
Red Hook, NY 12571

Some format issues inherent in the e-media version may also appear in this print version.

The Royal Institution of Naval Architects
8-9 Northumberland Street
London, WC2N 5DA
United Kingdom

Phone: 020 7235 4622
Fax: 020 7259 5912

publications@rina.org.uk

Additional copies of this publication are available from:

Curran Associates, Inc.
57 Morehouse Lane
Red Hook, NY 12571 USA
Phone: 845-758-0400
Fax: 845-758-2633
Email: curran@proceedings.com
Web: www.proceedings.com

International Conference on

Human Factors

19th-20th February 2020, London, UK

PAPERS

THE ROYAL INSTITUTION OF NAVAL ARCHITECTS
8-9 Northumberland Street, London, WC2N 5DA, Telephone: +44 (0)20 7235 4622

The Royal Institution of
Naval Architects

INTERNATIONAL CONFERENCE

Human Factors

19th -20th January 2020, London, UK

CONTENTS

***Unavailable at time of going to press**

'ONE HAND FOR THE SHIP': THE DEVELOPMENT OF ERGONOMICS GUIDANCE FOR ACCESS AND EGRESS

D. Oakley, University of Southampton, UK
J. Earthy and **A. Gibbins**, Lloyd's Register, UK

SUMMARY

Ships are hazardous working environments in which inadequate access and egress arrangements are a leading cause of injuries and fatalities [1]. There is growing awareness that ship design needs to consider the Human Element to mitigate risks to crew safety and Class Rules can contribute.

This paper outlines the development of ergonomics guidance for safe access and egress. It provides a background to the current regulatory environment and demonstrates the need for improved guidance on access and egress in the maritime industry. It outlines the development process including a technical justification of the scope, key criteria and application of ergonomics guidance for improved seafarer safety. It concludes with a critical analysis of the development process and a brief discussion on the challenges facing **rules ergonomists** [2].

Keywords: Safety, Occupational Health, Slips, Trips and Falls, Falls from Height, Maritime Ergonomics, Access and Egress, Regulation, Class.

NOMENCLATURE

ABS	American Bureau of Shipping
BV	Bureau Veritas
CSR	Common Structural Rules
ECL	Ergonomic Container Lashing
ESAE	Ergonomic Safe Access and Egress
GBS	Goal Based Standards
HEWG	Human Element Working Group
HCD	Human-Centred Design
IACS	International Association of Classification Societies
ILO	International Labour Organization
IMO	International Maritime Organization
ISO	International Organization for Standardization
LR	Lloyd's Register
MEPC	Marine Environment Protection Committee
MSC	Maritime Safety Committee
POLA	Principle of Least Astonishment
PPE	Personal Protective Equipment
RHFP	Rules Human Factors Panel
SOLAS	Safety of Life at Sea

1. INTRODUCTION

Access structures and access openings facilitate movement between and within the spaces onboard ship and are essential for effective working and living. However, a ship is a hazardous working environment and inadequate access and egress arrangements are recognised as a leading cause of injuries and fatalities amongst seafarers [1] [3] [4]. The pattern of reported accidents has remained consistent over many years and the hazards facing crews are well-understood [5]. The HE Alert! Bulletin [6] summarises a comprehensive list of principal hazards including:

- Slips attributed to inappropriate deck surfaces with poor slip resistance.
- Trips attributed to abrupt changes in deck level, loose fittings or low-level obstructions.
- Falls attributed to working at height, unstable surfaces, unguarded ladders, platforms or openings, or inadequate handholds and anchorage points.

The maritime industry is gradually putting more emphasis on the need for a safe and secure working environment and there is a growing awareness that the Human Element needs to be considered during ship design to mitigate risks to crew safety. Although, it has been identified that good ergonomics, safe working practices and the suitable provision of protective equipment results in greater awareness and an improved safety culture on ship [7]. The ergonomic design for seafarers is largely not well considered in the marine environment [8], and there is a distinct lack of knowledge in the integration of the human element and systems operational design amongst designers and other stakeholders. To meet increasing challenges for safety and performance, the link between ship design and ship operation has to be re-established and the marine industry must address Human Element concerns more explicitly.

2. BACKGROUND

2.1 INDUSTRY DRIVERS

International regulation has been increasingly responsive to the important influence of the Human Element on effective safety standards and practices, and has set out clear aims to improve working conditions for the seafarer community through better ship design. In 2004, the IMO restated its *Human Element Vision, Principles and Goals for the Organization* (Resolution A.947(23)) [9] including a specific goal to encourage the development of non-regulatory solutions and their assessment on the basis of human element principles.

There have also been a number of industry initiatives which identify safe access and egress as a particular concern. In 2006, the IMO Human Element Working Group (HEWG), convening at the Maritime Safety Committee (MSC) and Marine Environment Protection Committee (MEPC), noted that P&I Club statistics indicated that more than one out of five personal injury incidents reported were the results of slips, trips or falls [10]. The HEWG agreed that the consideration of ergonomics could result in a significant reduction of accidents in seafarers and subsequently approved an initial *Framework for Consideration of Ergonomics and Work Environment* (MSC-MEPC.7/Circ.3) [10]. The framework identified access structures such as stairs, ladders, ramps, walkways and work platforms as one of five key areas in which the application of ergonomics to design could reduce the incidence of personal injury and human error.

Additionally, in 2010, the IMO Maritime Safety Committee (MSC) adopted SOLAS regulation II-1/3-10 on *Goal-based Ship Construction Standards for Bulk Carriers and Oil Tankers* (GBS) (MSC.287(87)) [11] with a view to consider expansion to other ship types and areas of safety at a later date. The standard stipulated that the human element should be considered during structural design and arrangement with the overarching goals of safety of life, safety of ship and safety of the environment. GBS further outlined key functional requirements developed from experience, current rules and regulations and a systematic analysis of hazards and risks [12]. A specific functional requirement (II.9) illustrates the need to consider ergonomic principles to ensure safety during operations, inspection and maintenance, and specifies the provision of adequate means of access.

In response to the establishment of GBS, IACS developed the *Common Structural Rules for Double Hull Oil Tankers and Bulk Carriers* (CSR) [13]. The IACS Council was satisfied that these rules were based on sound technical grounds and achieved the goal of safer ships. However, CSR failed to set out explicit rules for the human element. IACS therefore investigated safe access and egress considerations through the development of the non-mandatory document *IACS Recommendation No. 132 – Human Element Recommendations for the Structural Design of Lighting, Ventilation, Vibration, Noise and Access and Egress Arrangements* [14]. Although the document summarises recommendations for the consideration of the Human Element and ergonomics in structural ship design and provides a broad means of compliance, a lack of unified interpretation has prompted unilateral action by Class to address the issue with the American Bureau of Shipping (ABS), Bureau Veritas (BV) and Lloyd's Register (LR) developing ergonomics rule sets.

2.2 LR'S RESPONSE

In response to the identified risks and hazards to seafarer safety and the increased industry drive to consider the Human Element within maritime regulation, LR has developed a strategy to address the Human Element. In 2003, an analysis of Class Rules and associated documents found a large number of implicit Human Element requirements [15]. In this analysis, Antonio and Earthy [15] concluded that in order to be of value, Human Element requirements in Class Rules need to be stated explicitly, consistently and to define the intended operational context in which they applied. Classification Rules and Regulations are the main means of mitigating error in design construction of ships [2]. Class Rules also provide a means of assurance and verification for the application of ergonomics. Hence, the inclusion of ergonomics requirements within Class Rules has the potential to make a credible impact and far-reaching improvements to safety on a large scale. The findings of the analysis resulted in the revision of rule requirements and the development of specific ergonomics rule sets [2].

The Ergonomic Container Lashing (ECL) notation was the first new physical ergonomics rule set to be approved by LR [8]. ECL was among the first rules products designed to mitigate against occupational health hazards, its primary goal to improve the safety of working arrangements for port workers and ship's crew when performing container securing, inspection and other related tasks. In developing ECL, Walker noted a particular challenge in translating from the task-based language of ergonomics to the prescriptive language of rules and requirements [8]. To ensure the development of criteria that are ergonomically sound, technically clear and assessable, Walker determined that ergonomics guidance must:

- Be prescriptive in order to provide clear unambiguous instructions to designers and manufacturers.
- Be defensible and of scientific merit in order to withstand challenges by stakeholders.
- Be written in simple language that is understood by non-ergonomists.
- Be verifiable such that compliance assessment is straightforward.

An awareness of the challenges faced in developing and gaining approval for the ECL notation provides a basis for the development of new ergonomics rule sets.

3. RULES DEVELOPMENT

In order to address the specific risks and hazards associated with access and egress, LR has developed a *Guidance Note for Ergonomic Safe Access and Egress.*

Class Rules communicate the requirements for minimum safe design to be satisfied for the consideration of Classification. However, there also exist Class products which offer optional requirements which Class can apply on behalf of an owner or shipyard as necessary. Notations stipulate additional requirements which enhance the minimum tolerable level of safety or provide assurance of performance in defined circumstances. Notations describe the ship and its appraised aspects and are appended to the character of Classification for the purpose of providing additional information of a ship's design and construction. On the other hand, Guidance Notes are complete documents which provide further explanation, clarification or technical recommendations on a specific topic supplementary to the Rules, their application may too be recognised by the award of a Notation.

A Guidance Note structure is adopted for the consideration of access and egress in order to best present this large body of material and provide a complete reference text and resource.

3.1 SCOPE

The scope of the Guidance Note is broad and aims to provide a 'whole ship' solution to access and egress design. Although a sweeping approach may risk industry resistance and places some restrictions on the level of description possible, it is adopted to ensure consistency is maintained across access and egress structures on a shipwide basis.

Design guidance is provided for a comprehensive range of commonly used access and egress arrangements on ship including:

- Access Structures – stairs, vertical ladders, inclined ladders, individual rung ladders, ramps, walkways, work platforms.
- Access Openings – doorways, hatches, manholes, scuttles.
- Access Control – doors, hatch covers, manhole covers.
- Safety Devices – handrails, guardrails, safety rails, safety cages, safety cables, fall arrest devices.
- Signage – safety signs, safety markings.

The Guidance Note stipulates that the design of permanent access structures and access openings should facilitate safe movement between or within the spaces onboard a ship. It should also allow suitable access for inspection, and maintenance tasks in spaces which are normally occupied, and inspection, survey and maintenance tasks in space which are not normally occupied. However, it should be noted that the guidance note does not provide explicit recommendations for emergency egress in the case of fire as this subject is extensively considered within the International Code for Fire Safety Systems [16].

The Guidance Note is not a re-examination of ship safety. It closely follows and extends some aspects of the design recommendations outlined in IACS Recommendation No.132 in order to maintain alignment with existing standards for access and egress. A significant amount of research was conducted in the development of IACS Recommendation No.132 and it provides a good basis for Rule development. The Guidance Note also complies with relevant statutory regulations except where it is deemed appropriate to raise the base line for safety on board ship by presenting more rigorous set of safety criteria.

The rationale for the consideration of ergonomics in the design of access and egress arrangements is based on the integration of safety and task performance. The Guidance Note considers physical ergonomics with recommendations for access structures, openings, control and safety devices designed for the expected range of anatomical, anthropometric, physiological and biomechanical characteristics of the ship's crew. The Guidance Note also considers cognitive ergonomics from the perspective of signage. Finally, the guidance addresses the context of its application and operational use, such that it encourages a shift in the perspective of design from the arrangement of structural elements to considering access structures as links between the spaces used by seafarers in working and living on the ship. From this perspective it is much easier to see the link between design and operational safety and performance. These contextual considerations are built into the guidance offering a 'multiplier' on ergonomics expertise for the designer.

3.2 SHIP TYPES

The Guidance Note presents general guidance for access and egress in order to provide a flexible accreditation route which will be applicable to a wide range of ship types. Cargo ships (32%) together with passenger ships (30.1%) are the main categories of ship where occupational accidents occur [1] and therefore represent a primary focus. The Guidance Note may also be applicable to special service craft, but is not currently deemed applicable to naval vessels or offshore vessels and installations due to the varied nature of their operational requirements.

3.3 STAKEHOLDER REQUIREMENTS

To ensure the use of the Guidance Note it is important to gain acceptance from both internal and external stakeholders. There are many stakeholders concerned with the safety of seafarers and ships in relation to access and egress. However, the links between them are often complex and variable requirements present a unique challenge to the development of ergonomics regulations and recommendations. The contractual environment in particular has not favoured consideration of human element issues in the past [17] due to increasing commercial pressure on both design and legislation. Key stakeholders include:

3.1 (a) Designers

Naval architects and marine engineers have a professional obligation to act in the best interest of seafarers to assure their health and safety [12]. Traditionally, naval architects addressed safety through the subjects of speed, strength and stability with the belief that adequate safety was achieved through compliance with Class Rules and Regulations. However, investigations into recent maritime accidents have highlighted the impact of the Human Element and ergonomics on safety level [12]. The problem arises where naval architects do not have the proficiency to analyse context of use or apply ergonomics principles and as such require precise recommendations and design rationale which do not require external expertise or interpretation.

3.1 (b) Shipyards

Shipyards are a critical stakeholder as this is often the body which chooses the Classification Society [8]. It is the shipyard's responsibility to ensure that a ship is built to the required specifications in line with regulation and Class. Shipyards are often reluctant to consider significant changes to current designs in case they increase the cost of build. However, ergonomics rule requirements do not need to be complex and can often be implemented (at least partially) through simple prescriptive rule sets to which will enable better quality assurance and improve efficiency long term.

3.1 (c) Owners

The ILO states that owners have a professional obligation provide and maintain ships in such a manner that, as far as is reasonably practicable, there is limited risk of accident or injury to seafarers [18]. The ILO [19] also emphasizes that ship owners should ensure that design of their ships takes account of ergonomic principles and conforms to relevant international and national laws, regulations, standards or codes of practice in order to achieve this goal. However, owners still need to be persuaded that the benefits of applying ergonomics recommendations outweigh initial and through-life costs. In addition to their professional and ethical obligations, it may be beneficial to emphasise to owners that there are financial implications associated with injury and fatality. The UK P&I Club has reported a rise in personal injury claims in recent years which has a direct impact on liability and premiums [3].

3.1 (d) Crew

Crews have a right to expect a safe and secure working environment which is operational and adequately protected from hazards [18]. Crews also have a right to expect that the ship will be well maintained throughout its lifetime to ensure safe working conditions are upheld [18]. However, crews must actively participate in ensuring safe working conditions are maintained. Crews are expected to take care of their own health and safety by following ship safety procedures and using the personal protective equipment (PPE) and clothing at their disposal [19]. Requirements also exist for inspecting, reporting and correcting unsafe conditions [19]. The NIOSH Hierarchy of Controls (Figure 1) [20] explains that over reliance on administrative controls and the provision of PPE are least effective in safety assurance. There is a requirement for hazard elimination, substitution or isolation via engineering controls which most easily implemented in the ship design phase.

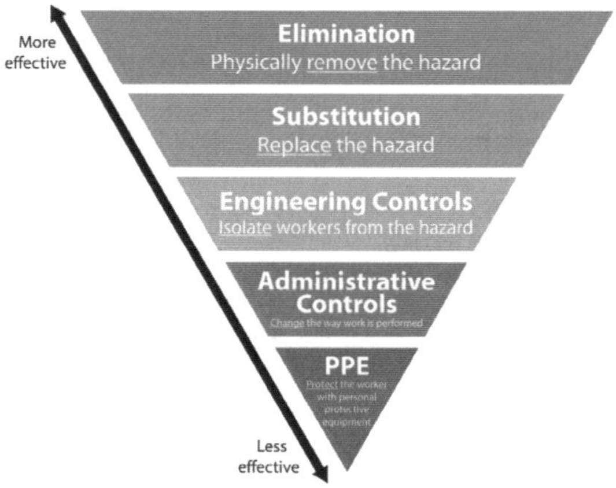

Figure 1: Hierarchy of Controls [20]

3.1 (e) Class Surveyors

Plan approval and field surveyors may be required to verify ergonomics requirements and recommendations. However, surveyors are trained to make judgements about engineering, not human behaviour [8]. Ergonomics recommendations that are not structurally defined cannot be easily verified on plans, hence, surveyors need to be provided with clear verifiable recommendations or alternate means of approval need to be made explicit. Additionally, if surveyors are expected to verify human element requirements and recommendations, a suitable training provision needs to be implemented to equip them with baseline knowledge of the subject [2].

3.1 (f) Regulators

Regulators stand to benefit from the success of non-mandatory frameworks such as a guidance notes or notations which can provide an evidence base or mandate for subsequent mandatory or statutory schemes to be implemented.

It is clear that the safety culture and economic pressures within stakeholder organisations have a significant effect on ship design, maintenance and operation [12]. The analysis of the stakeholder environment further illustrates the need for clear prescriptive guidance. Additionally, it highlights issues around the use of specialist terminology which can be misunderstood or vary within and between organisations and from ship to ship. All terminology used within the guidance note is consistent and clearly defined to ensure that essential safety issues are readily and reliably communicated both within and between organisations [12].

3.4 DESIGN CHALLENGES

A number of design challenges emerged during the development process of the Guidance Note. The following are some examples of specific challenges encountered in the development of ergonomic criteria for access and egress:

3.4 (a) Access Structures

The consistency of access structures plays a significant role in preventing accidents and injuries. For example, even small differences in the rise or going between adjacent steps can cause a misstep or fall. Consistency also plays a vital part in meeting user expectations. Human beings make mental models of everything of the they encounter and create patterns of behaviour based on interaction and experience [21]. If a structure meets expectations and corresponds to the mental model, there is more chance it will be used as intended and less chance of accident or injury. The Guidance Note therefore recommends consistency for the design of access and egress structures on a shipwide basis to increase usability and decrease instances of human error. It adopts a principle of least astonishment (POLA) which typically applies to user interface and software design [21] but has applications in structural design and significant implications for safety.

3.4 (b) Access Openings

The consideration of access openings presented a particular challenge to the development of recommendations for safe access and egress. The Guidance Note considers anthropometrics in the development of prescriptive recommendations to make provision for the range of variability anticipated in the expected user population. However, for the maritime sector, this becomes increasingly complex as the global nature of the industry is such that the expected user population is extremely variable. The employment of multi-national, multi-cultural and multi-lingual crews is commonplace [7] and the wide range of human variability and adaptability have led to conflicting practices worldwide. In particular, difficulties arise due to a failure to recognise differences in global anthropometry. Walker [8] observes this phenomenon in the development of ECL noting difficulties in '(convincing) designers in Asian yards for the need for sufficient headroom'.

Current specifications and locations for access openings, particularly for the purposes of inspection e.g. in double bottom, are insufficient for the majority of the global population, especially where persons are required to wear protective clothing or use breathing apparatus. The Guidance Note recommends adequate sizing for safe practical use.

3.4 (c) Access Control

The prevention of confined space deaths has been a growing concern for maritime industry. During the development of the Guidance Note, it was determined that good access control has a significant impact on crew safety, particularly in the prevention of entry into hazardous spaces. A confined space may be defined as a space which is not designed for continuous occupancy, where access is limited, where there is poor or no natural ventilation, and which may contain dangerous equipment or atmospheres [22]. However, in many cases confined spaces are not always easily identified and some have freely accessible openings for entry and exit with limited access control. For example, many hazardous spaces such as paint and chemical lockers, CO_2 and battery rooms can be entered through unlocked weathertight doors. The Guidance Note therefore recommends the use of appropriate access control devices in order to limit access into hazardous spaces including the provision of covers, mechanical closings, interlocks, and appropriate signage which makes clear any hazards and specified entry protocols.

3.4 (d) Fall Protection

The provision of fall protection arrangements is essential to prevent slips, trips and falls and in particular falls from height. However, fall arrest devices are typically used as first line of defence in fall protection. In contrast to this practice and acknowledging that this represents a change in the industry, the Guidance Note recommends fall elimination as the preferred method of fall protection with recommendations stating that working at height should be avoided where operationally possible. Guidance for fall protection arrangements and devices employs an adapted four-tier hierarchy of fall protection controls (Figure 2) for improved safety. This four-tier hierarchy of fall protection controls is categorised by:

- Fall elimination - the arrangements which facilitate the completion of required tasks without persons entering fall hazard areas.
- Fall prevention - the arrangements or devices which protect against the fall of persons including fencing.

- Fall restraint - the devices or personal protective equipment (PPE) which prevent the fall of persons including fall protections devices such as safety cables.
- Fall arrest - the devices or personal protective equipment (PPE) which provide the safe stopping of persons already falling including fall arrest devices.

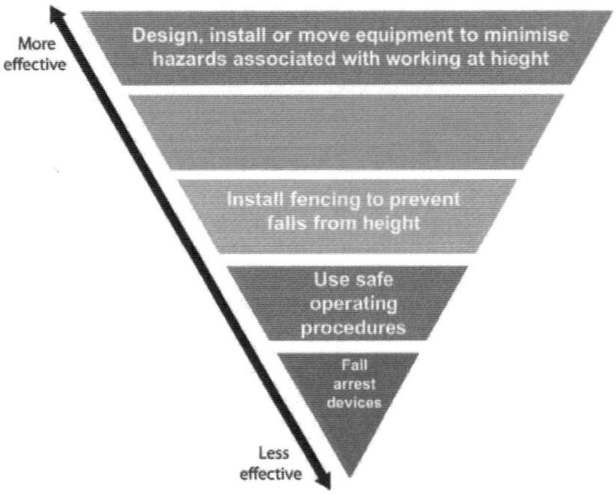

Figure 2: Hierarchy of Fall Protection Controls

It should also be noted that statutory requirements for fall prevention in ship design do exist within the International Convention on Load Lines (Regulation 25) [23]. However, this Convention only requires the provision of a low-level guardrails to prevent persons being swept overboard. In general, these requirements are sufficient for safe access and egress purposes, but, where fencing is in place for fall protection purposes the convention does not apply, and more appropriate ergonomic standards may be adopted. Fencing requirements to prevent falls from height are well-established for shore-based industry and there is strong evidence base for the incorporation of such requirements into the ship design process [2].

3.4 (e) Signage

The provision of guidance around safety signs and safety markings is a novel addition to the scope of the Guidance Note. It can be argued that for safety to be assured, the ergonomics of communications systems must also be considered including the provision of adequate signage in a language appropriate to the native languages of all onboard [7]. Recommendations for the provision of signage are provided within the Guidance Note with the purpose of facilitating the safe use of access structures, highlighting hazards to reduce the risk of slips, trips and falls, and controlling access to hazardous spaces. The Guidance Note presents a 'marinized tailoring' [17] of ISO standards for signage [24] that are widely applicable in other sectors and used effectively within terrestrial architectural practice. The Guidance Note also makes normative *ISO 24409 Ships and Marine Technology – Design, Location, and Use of Shipboard Safety Signs, Safety-related Signs, Safety Notices and Safety Markings* [25] which presents recommendations specific to the maritime industry, and provides guidance for its appropriate application in the context of ship access structures and arrangements.

3.5 APPLICATION

The *Guidance Note for Ergonomic Safe Access and Egress* provides an optional framework for best practice leading to enhanced ergonomics and improved safety. The accreditation of the Guidance Note is an important consideration from the perspective of uptake, as stakeholders will desire recognition for the appropriate application of ergonomics considerations. It is intended that the application of guidance note will result in the award of an optional Class Notation ESAE. It is also determined that survey requirements for assigning the ESAE Notation and for maintaining it though life will be fulfilled through the provision of a manual for safe access. The manual will be provided for initial plan approval and retained onboard for further reference at periodical survey to ensure compliance against the notation is maintained.

3.5 (a) Ship particulars and specification

Ship particulars and specification provides information of type of ship and identity.

3.5 (b) Access operational concept

The access operational concept is a verbal and graphic statement of design intent with regard to the operation of the ship. It is a user-oriented document that describes a designer's understanding of the users' needs and how a system will operate to fulfil those needs. The operational concept framework is typically used within systems engineering [26] and is an effective tool to establish a link between ship design and ship operation. The access operational concept should describe the form and use of access and egress arrangements with the ship 'system' and describe a strategy for access and egress through a statement of design rationale. In general, it should include:

- A statement of the goals, objectives and motivations of the systems together with the strategies, solutions, tactics, methods, and techniques used to accomplish them.
- A statement of the boundary conditions and constraints of the system. For example, there may exist a constraint on the hours of operation of the system, a constraint on the number of personnel available to operate the system or a constraint on operational facilities or space.
- A user profile including responsibilities, education, background, skill level, activities,

and modes of interaction with the current system.
- A description of the system including the operational environment and its characteristics, major system components and the interconnections and interfaces, performance characteristics and attributes, such as speed, frequency, flexibility, maintainability, portability, and usability and operational risk factors. Provisions for safety, security, privacy, integrity, and continuity of operations in emergencies.
- A graphical overview of the system. This can be in the form of any type of diagram that depicts the system and its environment.

3.5 (c) Ship general arrangement

The general arrangement gives an overview of the whole ship access and egress strategy. The current convention is to present the general arrangement of a ship via plan and profile views. However, it is counterintuitive to design access and egress requirements in plan and this practice is rarely seen in terrestrial architectural design. It is recommended that sectional views should be included as part of the general arrangement provision.

3.5 (d) Structural details

Plans and sections showing the permanent means of access to/from and within each space with appropriate technical specifications and dimensions.

3.5 (e) Inventory of permanent means of access

The inventory of permanent means of access should group means of access together into operational systems and include information on any features relevant to the ergonomic design of the access structure.

3.5 (f) Records of inspections and maintenance

The records of inspection and maintenance should include as a minimum the date of the inspection, the name and title of the inspector, the means of access inspected, verification of continued serviceable condition or details of any deterioration or substantial damage found [27]. This is vital for both ship safety through life and approval by Class.

3.5 (g) Records of change to permanent means of access

All changes to permanent means of access should be recorded. Major changes may require revision of the manual of safe access and will be subject to review and/or re-approval by Class.

The provision of a manual for safe access is in line with goal-based standards which stipulate design transparency [11], a specific functional requirement (II.10) states that ships shall be designed under a reliable, controlled and transparent process made accessible to the extent necessary to confirm the safety. The manual of safe access is a clear record of design intent and has subliminal human factors applications in providing a vital link between design requirements and operations. Additionally, in aligning ergonomic design principles with naval architecture standard practice and nomenclature in this way ensures human factors guidance is easily understandable and more credible in an operational framework.

3.6 CRITICAL EVALUATION

The Guidance Note was developed under a governance procedure that ensures the development of criteria that are ergonomically and technically sound, unambiguous, assessable and acceptable from a commercial point of view. Both the concept and the resulting work were reviewed by an independent, external body, the Lloyd's Register Rules Human Factors Panel (RHFP). The RHFP benefits from expert members from a wide range of marine industry stakeholders. The RHFP also designated mentors to provide advice to the project.

In order to identify and take account of requirements and feedback the project approach was iterative, and user centred. Rather than working from first principles the project derived technical guidance from international and industry standards. This material has already been through a rigorous review process and usually specifies the context in which the guidance is valid. In addition to the technical content, the user requirements and the document architecture were reviewed to validate value and user experience respectively. The first draft was reviewed by ergonomists and Rules experts, and following revision was submitted for formal technical review by internal and external experts in naval architecture, structural design and Class survey. This is the first formal step in the Lloyd's Register Rules approval process.

4. CONCLUSIONS

This paper presents an overview of the process of developing ergonomics guidance for safe access and egress. The paper illustrates many opportunities and challenges in the application of ergonomics within Class Rules. Addressing ergonomics through Class Rules has the potential to provide a wider benefit than consulting activities can provide. However, successful application of ergonomics requires working within the constraints of the maritime industry in order to make a tangible impact. Ergonomics continues to face scepticism from industry stakeholders who question its cost benefit; and the design community who question its value, reliability and validity [28]. Challenges also exist in raising human element competence for designers and surveyors [2].

It should also be recognised that the development of the Guidance Note is just the first step in operational safety. In order to achieve more effective safety assurance, designers, ship owners and operators will need to look beyond recommended arrangements and dimensions and adopt a goal-setting methodology to take account of hazards that are not adequately addressed through prescription [12]. This becomes increasingly important as new technologies emerge, and designers need to request derogations from Class. In cases of derogation, the designer is be required to present an evidence base to show that access and egress arrangements are safe. Lloyd's Register's development of procedural guidance on human-centred design (HCD) [17] [29] [30], is a good example of how guidance may be applied to novel designs to facilitate the integration of human factors for improved safety.

Additionally, the Guidance Note and associated Notation only goes one step towards a complete consideration of enhanced ergonomic design for improved safety. There is a case to be made for a broader notation to cover the full range of ambient environmental considerations covered by IACS Recommendation No. 132 including lighting, noise, ventilation and vibration. However, the successful application of the Guidance Note for Ergonomic Safe Access and Egress is a first step to secure wider industry interest.

5. ACKNOWLEDGEMENTS

The authors would like to thank the LR Rules Human Factors Panel for their support and feedback throughout the Rules development process. Additionally, Dhwani Oakley would like to thank LR Marine and Offshore for the opportunity to undertake this work.

The views expressed are those of the authors alone and do not necessarily represent the policy of University of Southampton or Lloyd's Register.

6. REFERENCES

1. EUROPEAN MARITIME SAFETY AGENCY, 'Annual Overview of Marine Casualties and Incidents 2018', *EMSA*, 2018.

2. WALKER, O., 'The Human Element Competency Required for Design Appraisal', *Human Factors in Ship Design and Operation, RINA*, 2011.

3. UK P&I CLUB, 'Risk Focus: Slips, Trips and Falls', *UK P&I Club*, 2012.

4. THE AMERICAN CLUB, AMERICAN BUREAU OF SHIPPING, and LAMAR UNIVERISTY, 'Maritime Slip, Trips, Falls and Lifting Injuries', *ABS Lamar University Mariner Safety Research Initiative Safety-Related Database, The American Club*, 2018.

5. SHERWOOD JONES, B., 'Designing Out Slips, Trips, and Falls is Not New, Difficult or Expensive', *Issue 17 Accidents, Alert! – The International Maritime Human Element Bulletin, The Nautical Institute,* 2008. Available at https://www.he-alert.org/index.cfm/bulletin/Accidents [Accessed 10 January 2020].

6. NAUTICAL INSTITUTE, 'Mitigating Slip, Trip and Fall Hazards', *Centre Spread 17, Alert! – The International Maritime Human Element Bulletin, The Nautical Institute,* 2008. Available at https://www.he-alert.org/en/centre-spreads.cfm [Accessed 10 January 2020].

7. SQUIRE, D., 'The Human Element in Shipping', *HE00350, Alert! – The International Maritime Human Element Bulletin, The Nautical Institute,* 2013. Available at https://www.he-alert.org/en/articles.cfm [Accessed 10 January 2020].

8. WALKER, O., EARTHY, J. V., SHERWOOD JONES, B. and TOZER, D., 'Safety Onboard Ship. A Case Study in the Transition from Science to Enforcement', *International Conference on Human Performance at Sea*, 2010.

9. INTERNATIONAL MARITIME ORGANIZATION, 'Resolution A.947(23) Human Element Vision, Principles and Goals for the Organization', *IMO*, 2004.

10. INTERNATIONAL MARITIME ORGANIZATION, 'MSC-MEPC.7/Circ.3 Framework for Consideration of Ergonomics and Work Environment', *IMO*, 2006.

11. INTERNATIONAL MARITIME ORGANIZATION, 'MSC.287(87) Adoption of the International Goal-Based Ship Construction Standards for Bulk Carriers and Oil Tankers', *IMO*, 2010.

12. ROYAL INSTITUTE OF NAVAL ARCHITECTS, 'Safety Guidance for Naval Architects', *RINA*, 2010.

13. INTERNATIONAL ASSOCIATION OF CLASSIFICATION SOCIETIES, 'Common Structural Rules for Bulk Carriers and Oil Tankers', *IACS*, 2014.

14. INTERNATIONAL ASSOCIATION OF CLASSIFICATION SOCIETIES,

'Recommendation No.132 Human Element Recommendations for the Structural Design of Lighting, Ventilation, Vibration, Noise and Access and Egress Arrangements', *IACS*, 2013.

15. ANTONIO, L. and EARTHY, J. V., 'The Human Element in Class Rules', *Human Factors in Ship Design and Operation, RINA*, 2003.

16. INTERNATIONAL MARITIME ORGANISATION, 'International Code for Fire Safety Systems', *IMO*, 2015.

17. LLOYD'S REGISTER, 'The Human-Centred Approach - A Best Practice Guide for Ship Designers', *Lloyd's Register*, 2014.

18. INTERNATIONAL LABOUR ORGANIZATION, 'Maritime Labour Convention', *ILO*, 2006.

19. INTERNATIONAL LABOUR ORGANIZATION, 'Accident Prevention on Board Ship and in Port', *ILO*, 1996.

20. NATIONAL INSTITUTE OF OCCUPATIONAL SAFTY AND HEALTH, 'Hierarchy of Controls', *NIOSH*, 2015. Available at https://www.cdc.gov/niosh/topics/hierarchy/default.html [Accessed 10 January 2020].

21. SALTZER, J. H., and KAASHOEK M. F., 'Principles of Computer Design – An Introduction' *Massachusetts Institute of Technology*, 2009.

22. INTERNATIONAL ASSOCIATION OF CLASSIFICATION SOCIETIES, 'Recommendation No. 72 Confined Space Safe Practice', *IACS*, 2013.

23. INTERNATIONAL MARITIME ORGANISATION, 'The International Convention on Load Lines', *IMO*, 2003

24. INTERNATIONAL ORGANIZATION FOR STANDARDIZATION, 'ISO 3864:2011 Graphical Symbols — Safety Colours and Safety Signs', *ISO*, 2016.

25. INTERNATIONAL ORGANIZATION FOR STANDARDIZATION, 'ISO 24409:2014 Ships and Marine Technology – Design, Location, and Use of Shipboard Safety Signs, Safety-related Signs, Safety Notices and Safety Markings', *ISO*, 2014.

26. INETRNATIONAL ORGANIZATION FOR STANDARDIZATION, INTERNATIONAL ELECTROTECHNOICAL COMMISSION and INSTITUTE OF ELECTRICAL AND ELECTRONICS ENGINEERS, 'Systems and Software Engineering - Requirements Engineering', *ISO, IEC, IEEE*, 2011.

27. INTERNATIONAL ASSOCIATION OF CLASSIFICATION SOCIETIES, Recommendation No. 90 Rev. 1, 'Ship Structure Access Manual', *IACS*, 2019.

28. STANTON, N. A., and YOUNG, M. A, 'Giving Ergonomics away? The Application of Ergonomics Methods by Novices', *Applied Ergonomics*, 2003.

29. LLOYD'S REGISTER, 'The Human Element - Best Practice for Ship Operators', Lloyd's Register, 2007.

30. LLOYD'S REGISTER, 'The Human-Centred Approach - A Best Practice Guide for Equipment Manufacturers', Lloyd's Register, 2014.

7. DISCLAIMER

8. AUTHORS BIOGRAPHY

Dhwani Oakley is a current student of Ship Science at the University of Southampton. Her previous experience includes a background in architectural design which has given her a unique perspective on user experience within naval architectural practice. The work reported in this paper is the product of a summer placement with Lloyd's Register Marine and Offshore.

Jonathan Earthy is the technical authority for HF in Lloyd's Register, including the R&D programme and the introduction of Ergonomics into Class. His technical speciality is the assurance of the quality of Human

Factors Integration and HCD, with special reference to advanced ICT. He has participated in standards development since the mid-1980s and is convenor of ISO TC159/SC4/WG6 HCD for interactive systems.

Andrew Gibbins is the Emergency Response Technical Lead for the Lloyd's Register's Ship Emergency Response Service. In his previous role as the Lloyd's Register Rule Architect he has planned, shaped and advised several Human Element focused products and worked to highlight the importance of the Human Element in shipping and Class Rules. He has also worked internationally to develop the IACS position to the Human Element and ensure compliance with the IMO's Goal-based Standards.

HUMAN FACTORS INTEGRATION IN SUBMARINE SPATIAL DESIGN

J. Clarke and A. Welch, BAE Systems Submarines, UK

SUMMARY
BAE Systems Submarines undertakes large scale, highly complex design programmes and this paper provides a summary of the work undertaken by the Human Factors Team to integrate their role into the spatial aspects of the design process in one such programme. The challenge is being addressed from four key directions i.e. Leadership & Infrastructure, Interdisciplinary Engagement, Supporting Processes and Tools & Techniques. This multidirectional approach has improved collaboration and consequently design outputs. Lessons learned relating to the approaches taken and outcomes inform Human Factors Integration Plans for future projects as well as improve relevant activities, especially the design and completion of relevant documentation. Some aspects of the work such as the formal non-compliance process are being applied to other work areas within Human Factors, have the potential to be rolled out across the wider business and may have applications in a diverse range of industries where programme complexity can make traceability of decision making and customer acceptance challenging.

1. INTRODUCTION

BAE Systems Submarines has a long history of delivering submarines for the Royal Navy and currently has multiple programmes in progress. Future generations of submarines are widely considered to be one of the world's most complex engineering challenges. Technological advances, threat changes, new methods of design and production are driving a class of submarines of completely new design. Consequently, the programme proffers work packages encompassing a diverse range of Human Factors (HF) practice.

HF has been considered a key contributor to the programme from the outset and as the programme progressed there has been a demand for their involvement to grow both in breadth and depth. Increasing maturity of the design meant an increased breadth of HF input requiring more detailed analysis as early assumptions are replaced with more definite knowledge and decisions. The issues addressed also become more complex in part due to the challenges presented by the design and programme constraints. One of the areas where this is most evident is spatial design work.

Possibly the most challenging aspect of designing, building, living and working on a submarine is the spatial constraint. The BAE Systems Submarines HF team have a significant number of staff dedicated to analysing all areas of the spatial design including general arrangement and boat outfit. Over the last five years the HF spatial team have increasingly sought to optimise their contribution to the field of work and have done so through interdisciplinary team engagement activities, improvements to process, capitalising on the opportunities provided by programme leaders and exploiting available tools and techniques for problem solving.

1.1 THE PROBLEM

HF professionals are sometimes viewed as coming from a very different 'world' in the engineering domain and conflicting views are common in such large scale, complex design programmes. These can be magnified if HF representatives are seen to be unrealistic in their expectations of what is achievable within the engineering constraints on the design, especially where pressures of increasing complexity and cost benefit come to bear. It is also true to say that despite the term "Human Factors" being commonly recognised amongst the engineering community there are large numbers within it that are unfamiliar with what the role actually entails or hold widespread misconceptions such as the belief that its focus is on the design aesthetic.

Submarines present particular spatial challenges that are further impacted by other considerations such as build process, structures, weight and noise signatures. The size and complexity of the design means that there are many constraints which HF have to work within; particularly physical structure (i.e. finite amount of space) and the sheer number of items that need to be on-board in order for the submarine to function. It is expected to house many thousands of electrical items, over hundreds of kilometres of cables and function as a home, workplace and recreation space for a crew of over 100.

HF are just one of many stakeholders in the design which means HF requirements and guidelines often need to be traded with other considerations to achieve the best all round solution. In practice this can mean a need to target the things that really matter most to HF representatives (essentials over desirables); in this case, operability and habitability (life on board). Within this context it is necessary for HF teams to provide a greater level of evidence to support the rationale for proposals and manage decisions that may diverge from their own recommendations, guidance and requirements. In the case of the latter there is an additional need to maintain records that reflect the context within HF assurance is provided.

The extensive lead time for the design, build, test and commissioning of a nuclear submarine is also a particular challenge. Decisions made to agree good design early on are commonly revisited and may be overturned as over subsequent years the spatial arrangement matures or equipment type may change, due to obsolescence or new requirements being generated that affect the choice of items used.

Learning from the experience of previous class boat programmes it was considered that early HF input would be a benefit. Achieving greater Human Factors Integration (HFI) requires early involvement and a strategy that looks beyond a formal requirement for HF involvement and/or 'sign off'. To achieve high value integration, a holistic approach to the array of influencing factors across a large organisation and its professionally disparate groups is required.

1.2 RESOLUTION OF THE PROBLEM

Over time a wide range of actions were taken to enhance HFI in all aspects of the team's work, a significant proportion of which for some years was focused on spatial design. These actions can be grouped into four main categories; Leadership & Infrastructure, Supporting Processes, Interdisciplinary Engagement and Tools & Techniques. The main improvements, benefits and challenges are here discussed in turn along with the benefits, challenges and key learning points as identified by the HF team's spatial leads.

2. LEADERSHIP & INFRASTRUCTURE

The Ministry of Defence (MoD) stipulated a requirement for HF integration and assurance from the outset of the programme. This in itself secured the embedding of HF in the work and provision of assurance for the design. This senior level leadership served to establish the baseline involvement of HF, ensuring that they were "in all the right rooms" and was essential in the early stages when the value of HF was still unclear to many or appeared to obstruct/slow progress.

Early in the design stage the programme adopted a philosophy which embedded all key stakeholders / disciplines in the design development ensuring all parties design considerations were heard. As part of this, for the purposes of allocating work the submarine was divided into large spatial 'areas', each one having its own dedicated team i.e. Major Area Team (MAT). Therefore every MAT has a HF representative who is involved in all relevant aspects of decision making for that area of the boat. The higher level formal requirement for HF integration and qualification places their requirements and guidance on an equal footing with others within this forum. Once everyone was accustomed to HF having a say in all aspects of the spatial design it became the norm to challenge situations where they have not been consulted or were absent from a relevant meeting. Fundamentally, having early HF input into the design has resulted in opportunities for HF to demonstrate their value by improving the design at a stage when there was less 'competition' for space and therefore less resistance. Consequentially by the time such decisions became more controversial there was greater awareness, acceptance and appreciation of HF input.

Overall, the mandate established HF involvement across the project and thus financial support ensued, enabling the acquisition of further resource, facilities and development of capability. This has laid the foundation on which to build and grow the HF service offering.

2.1 LEARNING FROM THE CHALLENGES

The backing of senior leaders was crucial in the early stages. Without the mandate for HF involvement, it may have taken significantly longer for HF to establish their place and value within the MAT. However, the HF leaders were cognisant of the fact that such assimilation of HF resource into the day to day work of the MATs had the potential to give rise to conflicting interests. Such situations could be managed through interactions with the wider HF team who could act as independent qualified specialist advisors.

More positively, an additional benefit of the team's success has been the demand for their expertise, which has consistently grown over the years in line with more widespread acceptance of the value of HF input.

3. SUPPORTING PROCESSES

The HF team developed a number of processes at a number of levels to assist with the HF Spatial development.

Whilst HF advice was most commonly consistent, as the team grew the large number of HF experts in one place could indeed lead to subtly different HF advice being given and therefore a simple spreadsheet was set up to record any ad hoc guidance provided thereby encouraging a coalescence of advice.

On a larger scale, HF requirements and guidance were created and designed to be accessed by designers in the first instance. At an "official" level, a Technical Department Instruction (TDI) was developed to make managing and escalating non-compliances a formal process. This introduced greater rigour and consistency into the way HF advise teams on spatial matters and formalised the follow up of identified issues.

The supporting HF processes support a 'right first time' culture that is driven by the business. This is exemplified by the Access Requirements & Guidelines document which facilitates a design process where the human operator is considered and designed for up front rather than HF advice or approval being sought later. In the case of the latter, an initial design is likely to have already been developed and would require costly re-work to change, or indeed may not even be possible to change.

In such cases there is an increased burden of proof placed upon HF for the changes they would wish to see; a need to justify the large associated costs with evidence of serious safety or usability concerns. By clearly setting out from the start what should be being aimed for, for an operable and habitable design, HF play a role as enablers of the 'right first time' philosophy

.

HF engineers have found these processes to be a motivator for the pursuit of high quality design and a trigger for instigating further assessment of more complex problems. Additionally, they improve consistency of design and design quality by minimising ambiguity and removing large swathes of subjective decision making. HF engineers have needed to be pragmatists and recognise that desired clearances and access requirements may often not be possible but, having the requirements and guidelines compels all stakeholders to strive to reach a best achievable position. Highlighted issues must be addressed when raised in an issues log / tracker which drives an action to be placed on an individual. Equally the creation of a record of non-compliances (which captures mitigation strategies, a decision on acceptability and a justification for that decision) formalises and drives a stronger commitment to the process of issue resolution. This has also resulted in a consistent audit trail of the whole decision making process and justifications for acceptance of non-compliances which are viewed by the customer. The MoD resource assigned to oversee HF areas of work has enabled effective engagement, particularly in the review of design challenges and non-compliances. The success of this process has prompted other teams across the programme to develop their own versions

2.1 LEARNING FROM THE CHALLENGES

Much was learned through the development of guidance and non-compliance processes. Document management software alone could not be relied upon to make compromises and decision making traceable so the aim was to avoid non-compliance stalling the design whilst still being thorough enough to provide a robust record of evidence. Working with the relevant stakeholders on a number of iterations helped to achieve this.

In the case of development of guidance and requirements, documents had to be brief and Requirements were only specified for highest priority and measurable features. Despite the multitude of benefits such formal advice can come with drawbacks which if left unmitigated can impact on HF capability building. Too much prescriptive guidance increases the risk of HF team members thinking less about the guidance they provide and not getting the experience of working out all the relevant rationale and evidence base for the guidance they are giving This is a drawback of working on one large scale project where the guidance is produced once and used for the life of the programme as opposed to HF teams working on multiple smaller scale programmes where guidance generation is more regularly required.

4. INTERDISCIPLINARY ENGAGEMENT

Eliciting feedback from the range of stakeholders during development and revision of the Access Requirements & Guidelines and the TDI helped increase ownership of them reducing the likelihood of resistance if they had to comply with them or use the non-compliance process.

As the programme progressed more attention was given to a wider engagement strategy with engineers who had ownership of the myriad of mechanical and electrical systems within the design (System Owners) as well as other key stakeholders such as MATs, Integrators, and Designers. Presentations were given outlining the role of HF and how we work which provided an opportunity to talk in greater detail about what they do, common pitfalls and myths etc. This activity prompted the HF team to ensure there was a consistent message delivered to all stakeholders. Creation of a more common understanding of the HF role and the thought processes that drive HF decisions gave greater weight and credibility to the HF messages. Whilst related cost savings are difficult to quantify, a noticeable avoidance of lengthy debates based on subjective opinions will have inevitably led to savings in project time and therefore money.

As the programme progresses through the stages of increasing design maturity, formal reviews take place where the customer and internal representatives jointly review the design and action plans that address related issues. HF have a key role supporting System Owners in ensuring all aspects of the design housed within their spatial areas are qualified, any non-compliances are being progressed as appropriate and there is an auditable trail for all decisions and actions related to these. This may require following up relevant tracked actions, assessing interfaces or performing bespoke Human – Machine Interface (HMI) assessments. These reviews are important and there is significant pressure on System Owners to demonstrate through extensive authorised documentation that everything that needs to be done, has been done, is being done in a timely fashion or as a minimum has a sound rationale for why it is not being done. The relationship between HF and System Owners in particular through these times is one of mutual support and benefit

More recently, the HF team have become involved in increasing awareness of HF to newer audiences. In a bid to tackle a shortage of designers, the organisation began to run a "Craft 2 Draft" programme, which allows existing staff from any area of the business the opportunity to apply for training to become a qualified designer. The team developed and deliver an "Introduction to Human Factors" session to these groups. In this very practical 'hands-on' session, designers at the very start of their training gain an understanding of the basic concepts of HF that are of specific interest to them so they have them in mind from the outset, or at least have a good understanding of the HF stance if challenged on HF grounds later in their work.

4.1 LEARNING FROM THE CHALLENGES

There were some initial concerns that HF were removing the ability to apply common sense but gradually stakeholders came to understand that there were wider contextual considerations being taken into account and there were means of managing situations where the documented 'rules' may not fit. The engagement strategy needed to resonate with the 'Pub Ergonomists' i.e. stakeholders who think the decisions are easy and just "common sense", often personalising the solution - "I could do it".

SMEs were sometimes more accepting of operator related issues than was deemed acceptable by the HF Team as they were used to being exposed to tolerating the conditions and less than optimal design and therefore accepted them as normal. This presented the risk of a HF challenge to the design not being taken seriously as "The operators say it's fine". At times such as this HF needed to make a case to the operators as well as the designers as to why they should aim for better and strive for optimum design.

5. TOOLS & TECHNIQUES

The HF team received training in a variety of tools and have access to a range of facilities that aid their assessments. In addition to traditional low fidelity aids such as paper prototypes and common drawing software the work involves extensive use of Computer Aided Design (CAD) tools. Use of dynamic mannequin software initially had limited applicability as it did not meet the anthropometric outlines of the Target Audience Description (TAD). Work was done to quantify the delta which de-risked its use but more recently the team has had access to more customisable mannequins which facilitate more accurate use.

5.1 PHYSICAL MOCK UPS

Where spatial assessment has been particularly enhanced is through the use of physical mock ups. An iterative facility houses full scale working mock ups of the boat control rooms which are primarily used for the development and testing of on board computer systems and related HCI/HMI but provide invaluable opportunities for performing spatial assessments. When considering what the facility is designed to do it can be it is regarded as relatively inexpensive, however the HF team's own Mock Up Facility is even more so.

The HF Mock Up facility is not a very large space but has sufficient footprint to house multiple to-scale mock ups at any one time. Measurements and screenshots from the CAD model are used to direct the build process with items usually built using foamboard. A large proportion the materials used can be later dismantled and reused. The mock ups created to support assessments are diverse (see Table 1).

Purpose	Example
User involvement in design	Early control room design
Spatial arrangement suitability for multiple users, traffic flow assessment	Ward Room, Messes
Acceptability of design change	Requirement to reduce compartment size
Supporting assessment of more complex non-compliances	Firefighting access through logistics trunk
Understanding clearances required for dynamic task sequences	Firefighting compartment re-entry procedure through doorways and hatches
Likelihood of local hazard to individuals	Protruding valves/equipment from bulkheads/ deckheads
Task performance and timing to support HRA	Reversionary throttle control

Table 1: Uses of physical mock ups

Where more robust structural integrity is required other mock up facilities are available to the team. Full scale wooden mock ups have enabled a number of trials assessing manoeuvrability and access. Other teams on the site have built and helped with assessments of other boat areas such as the appearance and comfort of the accommodation spaces and in-build welding access for more confined spaces such as the .bridge fin.

Physical mock ups are probably the most useful of all the techniques used by the spatial team as they enable decision making on more complex problems where on-paper/on-screen information has failed to persuade one way or another. In particular, such mock ups can provide information about dynamic tasks where natural body movements or users' physical behaviour cannot be seen or predicted through other methods. This is also the case where there is a need to assess tasks or space with more than one participant.

In addition to the technical benefits of physical mock ups they provide a medium through which to engage stakeholders, especially senior stakeholders. They enable them to get a first-hand look and interact with the space when they may not even have seen screen shots in the CAD model.

5.2 VISUALISATION

Use of immersive 3D visualisation using Head Mounted Displays in purpose built suites is routine. This Virtual Reality technology is particularly useful for assessing confined spaces and equipment dense areas. In such scenarios an appreciation can be formed rapidly of the constraints or likely acceptability of issues such as reach.

Augmented Reality facilities are available but at this point in time they have yet to prove useful in spatial assessment. Current thinking is that there may be greater potential to exploit this technology once the boat is further into build phase where it could aid in visualising the impact of design change requests.

5.3 LEARNING FROM THE CHALLENGES

There were long notice periods required to recruit participants to get involved in physical assessments and strict time limitations on the time slot available. If involving crew or clients the assessments needed to be thoroughly planned with clear goals and structure.

The most technologically advanced tools were not necessarily the best. It was more important to choose the right kind of technique depending on what exact information was required, or a combination of multiple techniques to draw on their differing advantages and compensate for individual limitations.

6. CONCLUSIONS

Early integration of HF in the design is crucial to establishing the value placed on their contribution by the organisation's leaders and gives you time to demonstrate credibility and value which builds trust. The latter is assisted by quantifying the benefits as best you can.

Of all of the challenges perhaps the greatest has been the conflict arising from the ongoing drive for decision making versus design maturity. The pressure of this had the potential to create frustration on all sides. Earlier in the design process many decisions had to be caveated due to the likelihood of future change. Later, HF amendments to design stemming from change to systems or equipment were harder to progress due to the costs of the change process and greater restrictions on available space. Every HF team member (regardless of their area of work) had to learn a higher level of pragmatism recognising that where programmes of work are so complex and the lead times so long, as design matures you often go from the optimum design to the optimum achievable compromise, balancing the need to stand by recommendations where it matters most; where the impact or consequences are greatest if not adopted.

Addressing improvements to HFI has not been a linear course with a clearly defined start point; rather a collection of decisions, mandates and deliberate actions that have been implemented and refined over time, each having an impact on HF input into the programme. Each of the four categories discussed here has contributed to positive outcomes. The leadership has ensured the embedding of HF and provision of assurance across the project. This support has also made funding available, enabling acquisition of resource and facilities and development of capability. This has laid the foundation on which to build the HF service offering.

The creation of a number of supporting processes introduced greater rigour into the way HF advise teams on spatial matters and formalised the follow up of identified issues. Human Factors engineers have found these processes to be a motivator for the pursuit of high quality design and a trigger for instigating further assessment of more complex problems. Wider stakeholder engagement is a diverse and ongoing process. Collaboration with the SMEs has added credibility to decisions made and a focus on engagement with influential spatial lead engineers has ensured important debate is allowed to take place.

Use of the wide variety of tools available facilitates a thorough approach to solving complex spatial problems and supports low cost redesign optioneering. It can also quickly demonstrate what things are not problematic which can provide closure on issues that might otherwise have continued, or intermittently re-emerged, to sap project time and resource.

The approach here has produced an array of benefits and has potential for wider implementation across the business and industry partners. Lessons learned illustrate methods for integrating Human Factors into complex design programmes, but also have applications for teams wishing to improve integration of their expertise into multidisciplinary projects as well as greater discipline and traceability in their decision making processes.

7. ACKNOWLEDGEMENTS

Thanks to Nicki McCann and Vicki Wall, Human Factors Specialists from the BAE Systems Submarines Human Factors Team for sharing their spatial assessment experiences for the purposes of this paper.

8. AUTHORS BIOGRAPHY

Julia Clarke is a Human Factors Specialist at BAE Systems Submarines. She is responsible for a range of Human Factors work including human error analysis and alarms and warnings management. Her previous experience includes optimisation of the boat Human-Computer Interface and supporting and leading spatial assessment activities.

Anna Welch is a Human Factors Specialist at BAE Systems Submarines. She is responsible for overseeing the activities of the Human Factors and Operability Teams working on platform and site related activities. Her previous experience includes conducting Human Factors assessment of operator interfaces to optimise the design for submarine crew.

IMPROVING HUMAN-CENTRED DESIGN APPLICATION IN THE MARITIME INDUSTRY – CHALLENGES AND OPPORTUNITIES

V. D. Vu and **M. H. Lützhöft,** Western Norway University of Applied Sciences, Norway

SUMMARY

Ergonomic issues with the design of ships and on-board equipment have been reported to hinder the crew's performance, increase the chance of erroneous actions and, consequently, lead to negative outcomes of maritime operations. One solution to such issues is following a human-centred approach in system design and evaluation, which helps improve usability. While being widely applied in other sectors such as in designing aviation systems, medical equipment, or consumer electronics, HCD practice is still limited within the maritime field.

Acknowledging this issue, the Nautical Institute organised a focus group with the participation of 78 people representing various maritime stakeholders. The participants jointly discussed and proposed solutions to improve HCD adoption in the maritime industry. Proposed solutions were aimed toward promoting awareness of usability and HCD among all stakeholders, increase HCD practices through regulatory documents, provide means to support designers in adopting HCD, and introducing a sector-based strategy to adopt HCD.

1. INTRODUCTION

The design of ships and shipboard systems directly influences the way people work onboard and has major impacts on outcomes of shipboard operations.

Improper designs can lead to negative consequences. For instance, an analysis of 66 fatalities in British shipping between 2003 and 2012 indicates that design flaws in ship structures can be connected to slips/trips/falls and enclosed space accidents, which accounted for 64% of the studied cases [1]. Certain design aspects of navigation equipment have been listed among contributory factors leading to accidents, as in the case of the Seastreak Wall Street, the Ovit, the Muros, and many others [2]. Results from other studies have clearly established that improper design remains a safety issue in the maritime industry, originating mainly from the lack of human factors consideration [3].

Traditionally, the design and development of technical systems has been dominantly technical-centred, aiming toward additional functions, more information, and increased automation while ignoring human abilities and limitations [4]. The operators are expected to follow required practices when cooperating with the machine to maintain proper system performance, only sometimes prepared and assisted by through training and education, To take, as an example, the case of the Seastreak Wall Street previously mentioned. The master was expected to be aware of the active mode of propulsion control, despite having to work with a control panel with almost-identical indicators for different modes [5]. Evidently, training and experience cannot in themselves overcome design flaws.

One solution that will reduce such issues is to consider the needs and abilities of users, and the context of use, when designing technologies. In other words, follow Human-Centred Design.

2. HUMAN-CENTRED DESIGN

Human-centred design (HCD) is a design approach that places user needs, abilities, and purposes at the centre of the design process. HCD is based on an explicit understanding of users, the tasks being performed and their working conditions. In HCD, the design process is interactive and iterative with design evaluation and modification being repeated to reach a design solution [6]. Figure 1 illustrates activities of an HCD process, according to ISO 9241-210:2010.

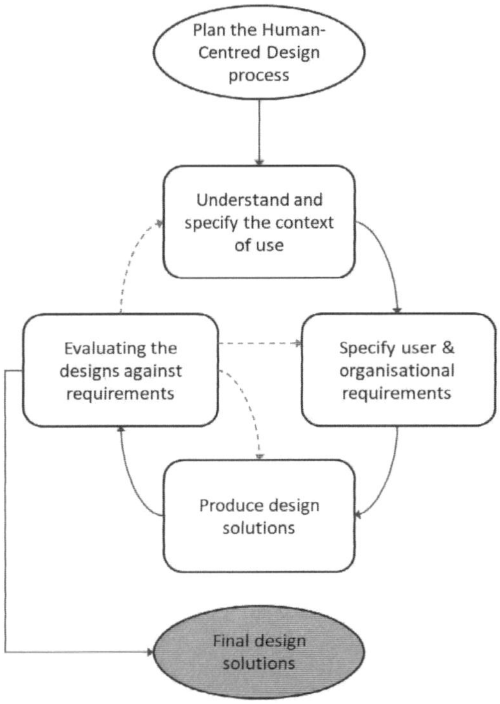

Figure 1: Human-Centred Design activities, according to ISO 9241-210:2010

HCD was put forward by Norman and Draper in the 1980s [7] and become popular in industrial and interaction design in the 1990s, especially after being advocated by the influential design company IDEO [8].

Following HCD will lead to products with high usability, which refers to the extent to which a product can be used by the intended users under the planned working condition to achieve the intended goals with effectiveness, efficiency, and satisfaction [6].

2.1. HUMAN-CENTRED DESIGN PRACTICE IN THE MARITIME FIELD

HCD has been implemented in several design projects of various scales in the maritime industry and there are efforts from the International Maritime Organisation (IMO) to promote such practice.

2.1 (a) Status

An example of HCD adoption in the maritime field is the design and construction of the Pure car, truck carrier (PCTC) Harvest Leader. The vessel was designed by Andreas Shipping Ltd. and constructed in Hyundai Mipo Dockyard for NYK Line. The design followed an unusual industry practice, in which the design team collected extensive feedback from crew members serving on other PTCT vessels to identify ergonomics aspects of other ships that can be avoided, improved, or, in the case of good solutions - followed. The feedback was analysed and forwarded to the shipyard, often in the form of proposed design solutions. The agreed-upon solutions were implemented in the final design of the ship. The end result was 1716 ergonomics improvements on a number of dimensions, including workability, habitability, maintainability, survivability, controllability, affordability, communication & user empowerment, and professional & psychosocial aspects [9]. Other examples of HCD practice in maritime include the "Platinum project" from SAM Electronics (now Wartsila), a user-centred bridge and engine control room suite [10], the design and construction of the Tamar-class lifeboats for the Royal National Lifeboat Institution, and the Ampelmann offshore personnel transfer system "walk to work" for Damen Service Operations Vessel 9020. Such cases demonstrate the positive outcomes of adopting HCD.

At the industry-wide level, the International Maritime Organisation (IMO) has recognised the need to consider user needs when developing future marine systems and has taken initiatives to promote HCD adoption. The e-Navigation initiative and particularly the Guideline on Software Quality Assurance and Human-Centred Design for e-Navigation are examples [11].

Despite the efforts of the IMO and individual stakeholders, it is a general view that HCD remains out of ordinary practice in designing ships and shipboard equipment. There is no systematic HCD application and the cases discussed above are not common. Maritime designers have been considering designs mainly from technical and economical viewpoints [12].

2.1 (b) Challenges

There are several reasons behind the current status of HCD adoption in the maritime field.

Firstly, the traditional mindset of maritime designers is technologically focussed rather than human-centred. Most naval architect students are not aware of or lack understanding of Human Factors and Human-Centred Design. These topics are also not commonly incorporated into educational programmes for future maritime designers [13].

Secondly, doing HCD is not simple. HCD is based on understanding users and iterative design. To obtain explicit knowledge of user needs and working conditions, and to translate them into design solutions requires skills from several domains, such as cognitive science, behavioural science, psychology, interaction design, etc. As a result, successful HCD requires successful coordination and cooperation of a multidisciplinary design team. Additionally, HCD is an iterative process. Doing HCD involves short-term risks to project time and cost but will lead to long-term benefits safety, efficiency, and also cost. As such, successful HCD requires support and consensus from all relevant stakeholders as well as determination and faith in the iterative process [12].

Thirdly, there is the monetary challenge. Doing HCD requires some additional expenses in the early design phases. However, there is currently no formal method for evaluating economic benefits to justify such investment, especially considering the fragmented structure of the maritime industry [12].

Finally, there is a lack of guidelines to support maritime designers in adopting HCD, specifically in designing digital user interfaces [14].

3. A FOCUS GROUP ON HCD ADOPTION IN MARITIME

Considering the current status of HCD adoption in the maritime field and the identified challenges, the Nautical Institute organised a focus group on the topic. The purpose was to identify solutions to promote HCD adoption in the industry.

The focus group comprised of 78 participants representing a wide range of stakeholders, including seafarers, ship managers, policymakers, academics, naval architects, manufacturers of marine electronics, and human factors specialists from other industries. Most participants were familiar with ergonomics issues in ships and shipboard equipment. The participants were provided information on the current status of HCD practice in the field and major challenges. Subsequently, the participants were divided into multidisciplinary teams and each team was required to discuss and brainstorm solutions to promote HCD.

The following sections will present the solutions proposed by the participants.

3.1 PROMOTE AWARENESS OF HUMAN FACTORS AND HUMAN-CENTRED DESIGN AMONG STAKEHOLDERS

Comments from the participants indicate that awareness of human factors and human-centred design has not been forwarded to all stakeholders. Participants commented that there is a need to promote industry-wide knowledge of such topics.

To specifically target seafarers and shipping companies, participants suggested taking advantage of the traditional reactive attitude in the industry and use accident investigation reports to inform ship owners, managers, and seafarers of the negative impacts that improper design can have on safety. Participants also proposed incorporating human factors knowledge into general training programmes for crews and safety personnel of shipping companies.

For the rest of the stakeholders, it was proposed that associations of seafarers, such as the Nautical Institute, should act as the moving force to promote HCD adoption and strengthen the request of seafarers for usable equipment. Such organisations should maintain influence at national, regional, and international levels.

Another proposed solution to promote industry-wide awareness of human factors and HCD is to hold regular meetings between stakeholders, similar to the workshop organised by the Nautical Institute.

3.2 THE REGULATORY APPROACH

The workshop also included participants from other industries, such as the railway and aviation industries. Through these external specialists, a number of solutions were proposed concerning regulatory changes that can affect HCD adoption in the maritime industry.

HCD has been widely adopted in the railway industry and there are regulatory documents requiring manufacturers to demonstrate that a system or product has undergone various design processes and that human factors have been properly considered [15]. A similar solution is proposed for the maritime industry, which would introduce an evaluation regime for ships and shipboard equipment that requires shipyards and equipment manufacturers to demonstrate human factors aspects considered in each design step.

Another solution is to introduce a similar solution to the LOSA (Line Operations Safety Audit) of the aviation industry to the maritime industry [16]. With LOSA, independent observers follow the work in the cockpit to observe the aircrew at work, paying attention to competency and procedures but also the design of flight equipment. The participants proposed to introduce similar solutions to the maritime industry. Once the independent observers notice design issues with shipboard systems that can inhibit usability, they can report the issues to designated organisations to bring it to the attention of manufacturers.

3.3 PROVIDE ADDITIONAL GUIDANCE TO SUPPORT HUMAN-CENTRED DESIGN PRACTICES

3.3 (a) Database of user feedback and design recommendations

As discussed in section 2.1 (b), successful HCD requires explicit understanding of users and contexts of use. Given the fragmented structure of the maritime industry, it is not simple for manufacturers of marine systems to study and understand user needs. One of the manufacturers of marine electronics commented that it was difficult to obtain user input, and response rates in usability studies using questionnaires were low, approximately 20%. The proposed solution was to introduce an online platform to connect end-users – seafarers with manufacturers. The seafarers can provide feedback and report usability issues through such channel and the information if made public so that all manufacturers can access. This solution already exists in the form of the Comité International Radio-Maritime (CIRM) User Feedback Forum. However, for several reasons it is problematic to get manufacturers to use this forum, including difficulties getting internal approval and the fact that some manufacturers have their own forums. Nevertheless, such a channel of communication between users and manufacturers could help promote HCD adoption and should be supported.

Manufacturers are encouraged to share design recommendations, and periodic meetings between industry stakeholders as discussed in section 3.1 can be a venue to distribute ideas of good design practices.

To address the issue of low response rates from end-users mentioned previously, participants representing seafarers suggested targeting training centres and maritime colleges as seafarers are more relaxed and more likely to respond when undergoing training at shore-based institutions.

Regarding the fragmented structure of the maritime industry, it was proposed that communication between seafarers and manufacturers should be made through representative organisations. For instance, the seafarers are represented by the Nautical Institute and manufacturers represented by the CIRM.

3.3 (b) Additional guidance to support HCD

Among the challenges to HCD adoption is the complexity of HCD itself and the additional expenses to perform HCD activities. A solution to improve HCD adoption is to provide maritime designers with means to lower both the difficulty of HCD activities and the expenses to perform such activities. The database of user feedback and design recommendations proposed in section 3.3 (a) is one example.

Another example of such means is the results of the OpenBridge project. The project introduces several 'standard' components of Graphical User Interfaces (GUIs) for navigation systems, such as toggle buttons, control panels, symbols and digits. These components are developed with usability in mind and manufacturers can adopt these components in the GUIs of their own systems without investing in developing new GUI features [17].

3.4 A SECTOR-BASED APPROACH TO HCD ADOPTION

In section 2.1 (a), we mentioned that there is currently no systemic application of HCD in the maritime field. One reason can be due to the large scope and fragmented structure of the maritime industry, making it difficult to start any initiative at the industry-wide level.

The design, construction, operation and safety requirements differ between ship types. As a result, the participants suggest promoting HCD adoption at a sector-based basis, starting with sectors with higher safety requirement such as the passenger cruise sector.

4. CONCLUSION

Human-Centred Design is an approach to system design that will lead to improved usability. Within the maritime industry, HCD has not been widely adopted due to the complexity of the approach itself and the lack of support from all relevant stakeholders.

The Nautical Institute hosted a focus group with the participant of 78 participants representing a wide range of stakeholders in the maritime industry, the purpose of which is to propose solutions to improve HCD adoption in ships and equipment design.

The proposed solutions fall into four categories:

- Promoting awareness among stakeholders of the benefits of HCD adoption by targeting general training programme of shipping companies and holding regular meetings between stakeholders.
- Introducing regulations to make usability a criterion of successful system design.
- Provide means to lower the complexity and cost of conducting HCD. Such means include a database of user feedback, design guidelines and recommendations, and similar supporting tools.
- Promoting HCD adoption at a sector-based basis instead of aiming directly at an industry-wide adoption.

5. ACKNOWLEDGEMENTS

The authors are grateful to the Nautical Institute to organise the workshop and taking a lead in promoting human factors and human-centred design knowledge in the maritime industry. We also thank Inmarsat for providing the venue for the event. Above all, we thank representatives of industry stakeholders for taking part in the discussion.

6. REFERENCES

1. Sherwood Jones, B., *Regulation Of Occupational Health And Safety - A Human Factors Viewpoint*, in *SEAHORSE International Conference on Maritime Safety and Human Factors*. 2016: Glasgow, UK.
2. Lutzhoft, M. and V.D. Vu, *Design for safety*, in *Managing Maritime Safety*, H. Oltedal and M. Lutzhoft, Editors. 2018, Routledge: London. p. 106-140.
3. Grech, M. and M. Lutzhoft. *Challenges and opportunities in user centric shipping: developing a human centred design approach for navigation systems*. in *Proceedings of the 28th Australian Conference on Computer-Human Interaction*. 2016. ACM.
4. Billings, C.E., *Human-centered aircraft automation: A concept and guidelines*. 1991.
5. NTSB, *Allision of the Passenger Vessel Seastreak Wall Street with Pier 11, Lower Manhattan, New York, New York, January 9, 2013*. 2014: National Transportation Safety Board.
6. ISO, *9241-210:2010 - Ergonomics of human system interaction - Part 210: Human-centred design for interactive systems*. 2010, International Organization for Standardisation.
7. Norman, D., *Cognitive Engineering*, in *User centered system design - New Perspectives on Human-Computer Interaction*, D. Norman and S. Draper, Editors. 1986, Lawrence Erlbaum Associates, Inc.: Hillsdale, New Jersey, US.
8. Koskinen, I., et al., *Design research through practice: From the lab, field, and showroom*. 2011: Elsevier.
9. Bialystocki, N., *Human Factors Dimensions in the Design of a PCTC Vessel*, in *International Conference on Human Factors*. 2016, The Royal Institution of Naval Architects: London.
10. Wartsila. *Wärtsilä NACOS Platinum - More in command than ever*. 2019; Available from: https://www.wartsila.com/nacos-platinum.

11. IMO, *MSC.1/ Circ.1512 Guideline on software quality assurance and human-centred design for e-Navigation*. 2015, International Maritime Organisation.
12. Petersen, E.S., *User Centered Design Methods must also be User Centered: A Single Voice from the Field. A Study of User Centered Design in Practice*. 2010.
13. Abeysiriwardhane, A., *Shaping ships for people: human centred design knowledge into maritime education*. 2017, University of Tasmania.
14. Mallam, S.C. and K. Nordby, *Assessment of Current Maritime Bridge Design Regulations and Guidance*. 2018, The Oslo School of Architecture and Design: Oslo, Norway.
15. Network Rail, *Network Rail (Infrastructure) Ltd Health & Safety Management System*. 2018, London, UK: Network Rail (Infrastructure) Ltd.
16. Klinect, J., et al. *Line operations safety audit (LOSA): Definition and operating characteristics*. in *Proceedings of the 12th International Symposium on Aviation Psychology*. 2003. Ohio State University Dayton, OH.
17. Nordby, K., S.C. Mallam, and M. Lützhöft, *Open user interface architecture for digital multivendor ship bridge systems*. WMU Journal of Maritime Affairs, 2019. **18**(2): p. 297-318.

THE OPERATOR'S STAKE IN SHORE CONTROL CENTRE DESIGN: A STAKEHOLDER ANALYSIS FOR AUTONOMOUS SHIPS

E. Veitch, NTNU, Norway
A. Hynnekleiv and **M. Lützhöft,** HVL, Norway

SUMMARY

Technology driven innovation has been advancing the development of autonomous ships and their Shore Control Center (SCC) counterpart. We plot a stakeholder map focused on the SCC based on data collected in a workshop and semi-structured interviews containing ten (10) Subject Matter Experts. The results identify key stakeholders and their interconnected relationships. Furthermore, insights from the qualitative data highlight important human factor issues of SCC operators that can be put into organizational context using the stakeholder map. The contribution is a tool designers can use to understand of the business model and the needs of the SCC customer, technology developer, and operators. Considering the operator's stake in autonomous shipping and the human factors issues that are expected in the SCC, the case is made for a Human-Centered Design (HCD) process to run concurrently with the technology innovation.

NOMENCLATURE

AUV	Autonomous Underwater Vehicle
HCD	Human-Centered Design
IMO	International Maritime Organization
LOA	Level of Autonomy
MASS	Maritime Autonomous Surface Ship
OOTL	Out-of-the-Loop
ROV	Remotely Operated Vehicle
SCC	Shore Control Centre
STCW	Standards of Training, Certification and Watchkeeping
VTS	Vehicle Traffic Service

1. INTRODUCTION

The purpose of this work is to contribute to design methodologies for development of Shore Control Centres (SCCs). SCCs are control rooms where remote monitoring, supervision, and intervention of autonomous ships are coordinated via human operators. The paper is organized into three parts:

- Presenting a stakeholder map of the SCC based on data collected from an expert workshop and multiple interviews (Section 4: Findings)
- Uncovering human factors issue latent in the SCC concept based on the qualitative data (Section 4: Findings)
- Discussing how Human-Centred Design (HCD) and an understanding of key stakeholder relationships can contribute to design of human-automation systems in the SCC (Section 5: Discussion)

The original research presented in this paper is based on data collected in the HUMANE project, whose mission is to enable autonomous shipping through addressing human factors issues [1]. This paper is a result of a collaboration between ongoing doctoral research at NTNU (Norwegian University of Science and Technology) and HVL (Western Norway University of Applied Sciences).

2. BACKGROUND

2.1 AUTONOMOUS SHIPS & THE SCC

In 2018 the IMO formally distinguished MASS (Maritime Autonomous Surface Ships) from other types of self-driving or remotely-controlled marine vehicles, like AUVs (Autonomous Underwater Vehicles) or ROVs (Remotely Operated Vehicles) [2]. This announcement from the highest authority in international shipping addresses a contemporary technology push for automated shipping. Maritime industry leaders have already developed concepts for automatic crossing and docking of passenger ferries [3] as well as tugs operated remotely [4]. A partnership between a major ship management company and a technology supplier in Norway have signalled plans to operate the world's first autonomous cargo vessel in Southern Norway [5]. Governments, too, have adopted the concept of autonomous ships in strategic planning. The National Transport Plan of Norway recognizes autonomous transport as a transitional force in coastal shipping, with areas like the Trondheimsfjord being sanctioned as official test areas for development of autonomous ship technology, regulation, and shore infrastructure [6].
Fully autonomous ships do not exist yet. This means that for the foreseeable future, a human operator must in some way be "in the loop," supervising the operation and on stand-by to take over control in case of system failure. This work is envisioned to take place in the SCC. The need for human operators is also underscored in legal, regulatory, and insurance frameworks that place responsibility and liability onto the shoulders of individuals – not computer software.
Autonomous ships, then, are autonomous *with an asterisk* – they rely upon human supervision, control, and responsibility. Even with backup systems for control and redundancy in communication channels, it is likely that SCC operators will remain a safety-critical element of autonomous shipping. The early development of autonomous ships may thus have been affected by the so-called "substitution myth" – that is, the tendency for

developers to assume that new automation solutions are viable substitutions of human activities [7].

2.2 LEVELS OF AUTONOMY

Levels of Autonomy (LOA) for autonomous ships refer to discrete levels of navigation control that range from full human-operated control to full algorithmic control. Many LOA scales have been proposed to date [8]. For designers a challenge is synthesizing the various LOA definitions into a single working definition for autonomous ships. Rødseth et al [9] addressed this problem and suggested using the SAE J3016 standard for autonomous cars [10] as a working definition for ships. The Norwegian Forum for Autonomous Ships (NFAS), currently the largest professional society for autonomous shipping worldwide, has also endorsed the SAE standard for adaptation to autonomous shipping [6]. This standard describes a stepwise scale from 0 (No Automation), to 5 (Full Automation). Most industry experts agree that Level 3 (Conditional Automation) and Level 4 (High Automation), represent the highest levels of automation we can expect to be implemented onboard ship systems in the foreseeable future. The framework implies a transition to autonomy through increasingly automated and integrated control functions.

2.3 HUMAN FACTORS

This paper focusses on "human factors from the neck up," or "engineering psychology" [12] as opposed to more classical elements of ergonomics often related to fatigue and safe work practices. The work asserts that a good understanding of human factors issues may provide an important tool for design teams. Given the central role that humans are predicted to have in day-to-day operations of autonomous ship systems, knowledge of human factors issues can inform HCD practices, an approach some have called "human-centered autonomation" [13].

Section 4 (Findings) lists three high-level human factors issues that were central themes in the workshop and interview data. Section 5 (Discussion) examines these human factors issues in light of current theory and frames them in the organizational context of the stakeholder value chain.

3. METHOD

3.1 DATA

The data are transcripts obtained during a focus-group workshop and three interviews in the HUMANE project [1]. The HUMANE project has planned four major workshops about autonomous shipping; three have been completed to date. The mission of the HUMANE project is to contribute to the understanding of the human element in autonomous shipping. This paper is written independently of the HUMANE project group but uses the project's results from the first two workshops to accomplish a similar purpose. The difference is that this paper frames the SCC at the core of autonomous ship operations and analyses the qualitative data to parse out information that identifies the SCC's key stakeholders. It also has a secondary aim of identifying the roles of SCC operators and the associated human factors issues expected on the job. The HUMANE project scope was much broader that this theme; notwithstanding, the datasets were found to be highly relevant.

The data was separated into two categories: focus-group workshop and semi-structured interviews. The former was held with subject-matter experts (SMEs) on the theme of technology in autonomous shipping; the latter was about topics related to legal, classification, and regulatory implications. Table 1 and Table 2 present an overview of the data and characterize the main topics discussed. Interviews 1, 2, and 3 were similar because they followed the same semi-structured interview format.

Table 1: Characterization of the data

ID*	**Theme**	**Topics**	**Date**
WS1	Technology	Unmanned bridge & engine room, low-manning, shore monitoring, full autonomy, business case	10/2018
Int1	Regulation	Current frameworks, governing bodies, roles and responsibilities, business case	4/2019
Int2	Legal implications	IMO & STCW amendment needs, training, liability	4/2019
Int3	Regulation	Regulatory issues, human factors, testing and verification needs	4/2019

* WS = Workshop; Int = Interview

Table 2: Data metrics

ID	**No. SMEs**	**No. inter-viewers**	**Word count**	**Record length**
WS1	6	2	19,700	-
Int1	1	2	16,800	4 hrs
Int2	3	2	22,800	3 hrs
Int3	3*	2	28,000	3 hrs

* Contained same experts as previous interviews

3.2 PARTICIPANTS

Participants were selected by convenience and snow-ball sampling through the HUMANE project leaders' own networks. They were selected based on expertise and represented governmental organizations, classification societies, shipping companies, technology manufacturers, universities, and research institutions in the maritime domain.

In total there were 10 participants in the combined data sets of workshops (6) and interviews (4). Of the four individuals interviewed, three were interviewed in a second session (Int3); these interviewees are not double-counted.

3.3 ANALYSIS

The workshop and interview transcript analysis yielded insights relevant to SCC design. Specific items of interest were the identification of stakeholders in autonomous shipping and the various human factors issues associated with the work envisioned at the SCC. For the former, this included descriptions about how stakeholders were interconnected and about how they were distinct from traditional models of shipping. For the latter, this includes the ways in which human factors issues were envisioned to play out as safety-critical elements during demanding operations.
The main author independently analyzed the results from the HUMANE group researchers who had originally collected the data. Each data set was treated as an individual case study and was analyzed using cross-case synthesis [14]. Cross-case patterns were identified, aggregated, and finally generalized using argumentative interpretation in Section 4 (Findings). The findings were verified by the second author who was present at all interviews and familiar with the cases. Rival interpretations to the data analysis were considered and stem primarily from investigator bias, or reflexivity of the researchers in the field.

4. FINDINGS

4.1 STAKEHOLDER MAP

The purpose of the stakeholder map is to provide a visual tool to help design teams identify relevant parties who are involved in, or affected by, the operation of an SCC. As we will see in Section 5, this provides insights into HCD approaches. All stakeholders shown in the visual graphics were explicitly mentioned by the workshop or interview participants when they were describing a future autonomous shipping organization and the various parties that will affect them. The map represents an interpretation of the aggregate of those descriptions. The decision to place the SCC at the center was deliberate and reflects the focus of this paper.
The findings do not represent an exhaustive list of all relevant stakeholders. It does not even capture a single agreed-upon model for an SCC organization; interviewees were not clear, for example, whether the SCC would be a standalone company or whether it would a consortium representing ship managers, owners, and technology suppliers. It was generally agreed, though, that the SCC had an interest in close collaboration with a technology supply company and in operating ships on behalf of multiple customers. To underscore the importance of this fundamental set-up, a core stakeholder map is depicted as shown in Figure 1.

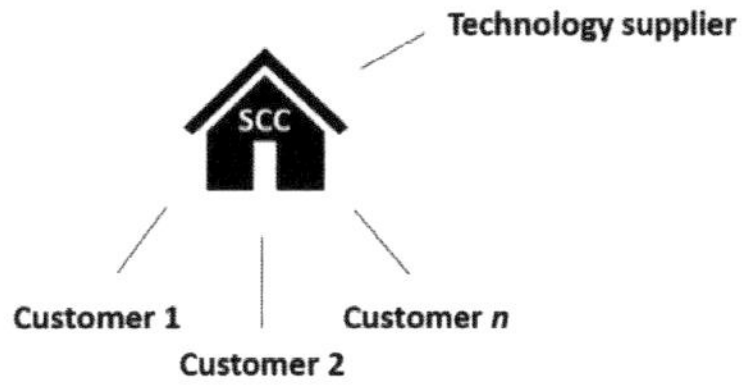

Figure 1: Core stakeholders of SCC company

In Figure 1, *n* represents *n* number of ship owners with separate interests in engaging the SCC to coordinate its autonomous vessels.
In Figure 2, the remaining identified stakeholders are included. They are connected with lines that represent interlinking relationships – either through exchange of value or information or simply by virtue of an acknowledged partnership. The stakeholders listed in the upper box in Figure 2 represent those that influence all others in the lower box in a less tangible way; like, for instance, through governing legislation, design guidance, or even public values.

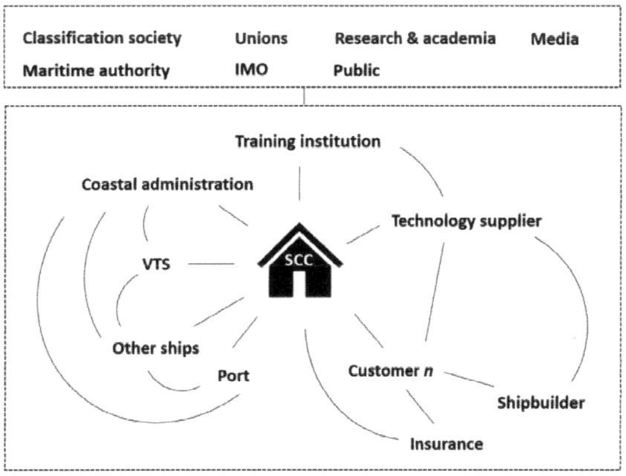

Figure 2: Stakeholder map of the SCC

The stakeholder map in Figure 2 is distinct from that of a traditional ship ownership ecosystem is that the ships are part of a larger transportation system and are not the primary investment in themselves. Interviewees expressed visions of relatively short sea routes that run on predictable schedules, around the clock, and at slow speeds relatively to manned ships. The value-added proposition stemmed from owning and controlling an integral part of logistics supply chain. In this vein, one interviewee stated:
"What I see is that the autonomous ships probably won't come from traditional ship owners. To a large degree they'll come from [companies like] Amazon."
With the ship framed as a single unit in a wider distribution network, it shifted the focus of future owners to the cost-effective and predictable delivery of cargo between two ports. The introduction of short-sea autonomous shipping, it was said of an interviewee would *"break up shipping lines into bilateral transport."* The advantage for the company is that they could

improve transportation reliability and not have to rely upon freighting on often congested road infrastructure.

4.2 STAKEHOLDER VALUE CHAIN

The stakeholder value chain is a stakeholder map that shows the direction of value (i.e. money) between stakeholders (Figure 3). Once again, the aim was to analytically interpret the workshop and interview data to build an aggregate picture of the flow of value in the envisioned stakeholder map around the SCC. Only those directions of flow that were identified in the data are included in Figure 3.

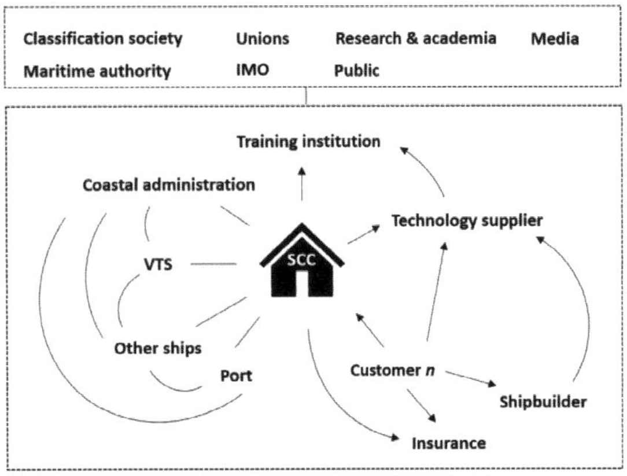

Figure 3: Stakeholder value chain of the SCC

4.3 HUMAN FACTORS

This section identifies three high-level human factors issues from the workshop and interview data. The three areas were chosen based on the frequency and duration they were discussed in the independent cases. The topics all represent known human factors issues related to automation and human performance. Note that although human factors was not explicitly a planned topic according to the workshop topics and semi-structured interview questions, it merited attention in all cases studies.

4.3 (a) Multitasking

It emerged from the results that the business case of autonomous ship operations from an SCC is closely linked to the number of customers, or ships, one SCC can manage (Figure 1). The obvious question, then, was "How many ships can one operator, or one team of operators, manage simultaneously?"
The tasks of the operator were characterized as a time-critical management of effort that varied according to the demand on the operator's cognitive resources. Influences that affected this demand were Level of Automation (LOA), similarity of tasks, and priority of tasks. For example, multiple vessels at a high LOA, all requiring minimal priority, would intuitively free up more cognitive resources than multiple vessels at different LOAs, with some requiring more priority than others. Examples for the former were open-sea sailing; for the latter they were docking or dealing with an emergency.
Multitasking proficiency was considered important given the imagined working environment that involved the risk for task overload, degraded hazard detection, and interference of ongoing tasks by alarms. Especially problematic was the task of remote manual control; at a time when it is needed it is unlikely that the operator will be able to focus on anything else. This fits with some researchers' models for multitasking which posit that humans are more adept at switching between individual tasks than performing multiple tasks simultaneously, and that this task switching requires cognitive effort [15]. In the case of manual control in the SCC, an unknown amount of time will be needed for this effortful task switching – time that may be safety critical.

4.3 (b) Situation Awareness

Considerable concern was expressed about the operators' capacity to maintain adequate situational awareness. Interestingly, despite the importance placed on this issue throughout the various case studies, no definition of situation awareness was provided, and it was unclear exactly what was meant by it. In this paper, Endsley's definition is used: namely, situation awareness consists of *perception* of critical elements in the environment, *comprehension* of their meaning, and *projection* of their status into the future [16].
Moreover, the term "Out-of-the-Loop" (OOTL) phenomenon was used to describe the tangible situation in which an operator's full attention would be suddenly needed, resulting a time-critical and stressful demand for cognitive resources to gain situation awareness. One example is an emergency manual take-over due to sensor malfunction; the operator would be expected to switch suddenly from a low- to high-demand task smoothly and effectively.
A related issue is the task of supervising highly automated ships without succumbing to boredom. In other words, the demands on the operator's resources were occasionally envisioned to be so low that it would quickly undermine their engagement, potentially exacerbating the OOTL phenomenon.

4.3 (c) Skill definition, acquisition and retention

A common theme among the case studies was operator skills; namely, what skills would be required, how they would be obtained, and how they would be retained. One area of contention that emerged from the case study results was that of what skills would be necessary to be a SCC operator. Most asserted that a certified seafaring navigator would hold the necessary credentials; however, some suggested that given the new working environment, a unique set of skills must be defined. One interviewee compared the situation to the unique skills required of ROV technicians, since they *"are trained in looking at a computer screen...as opposed to a mariner who needs to look out the window."*

The nature of the SCC work environment was also linked to de-skilling. One interviewee suggested that seafarers would lose their "ship sense," or "feeling of how a ship moves," and therefore their skillful application of decision-making dependent upon such an information channel would be compromised. On the other hand, another pointed out that software technicians hired in an SCC would likely not have the seamanship required to safely operate a ship in difficult weather conditions.

5. DISCUSSION

This section discusses how we may design for effective automation in shipping systems using knowledge of stakeholder inter-relationships and human factors.

Donaldsen et al [17] have shown that stakeholder value analysis (or "Customer Value Chain Analysis" as they call it) should be among the first steps in the design process, arguing that design flaws are often the result of poor understanding of the stakeholders and the flow of value and information between them. By understanding the key stakeholders, the design team can define the SCC in terms of the company's market and project objectives, as well as the needs of the ships' owners and the operators employed in the SCC.

To date, no SCCs are in operation. They are, however, in a sufficiently early stage of development to allow for this type of analysis to be used to its full advantage. As Buxton [18] has explained, design comes before engineering in a typical commercial product development process and should be focused on converging design requirements (Figure 4).

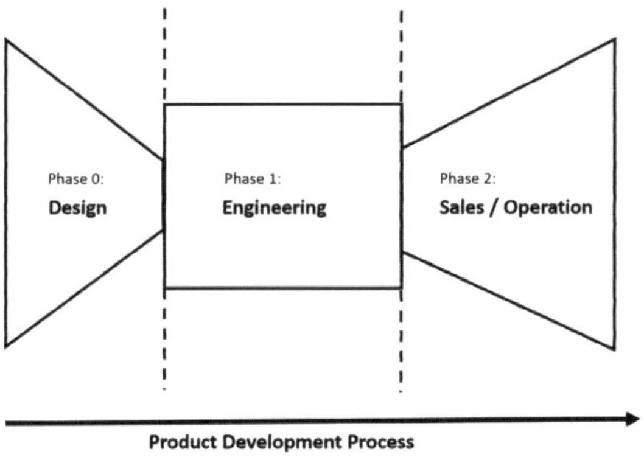

Figure 4: Stages of product development (adapted from Buxton [18]).

As Norman [19] has elucidated, often the customer is not the end user, yet in these cases implementing HCD is just as important as in the case when the customer is the end user. In the case of the SCC the customer is either a ship owner or a standalone autonomous shipping company representing the owner; the end users are the operators employed in it. To illustrate this with an example, consider the design issues that might result from accounting for just the customer's need (for example, in maximizing number of ships operated per operator). The result might be complaints from the operators about usability (multitask overload). The designer's role here is therefore to synthesize the feedback from the operators in a way that also fits the true customer's needs; the customer would save cost in this case by avoiding wasted time, training, and even lower staff moral and productivity stemming from poor usability and multi-task overload.

Such a feedback cycle can be represented graphically with the aid of the stakeholder map. For instance, in Figure 5, exchange of complaints and value are plotted among critical stakeholder following the analytical technique suggested by Donaldsen et al [17]. Based on the findings (Section 4), one plausible scenario may play out as follows: the SCC operator complains to the technology supplier about a high frequency of false alarms on a vessel's sensor. Meanwhile, the customer complains to the SCC that its cargo is delayed due to the extra time required to address the alarms. The role of the design team in this situation may be to facilitate the feedback mechanism to the technology supplier, speeding up the feedback process and effectively reducing disruptions to the customer's schedule.

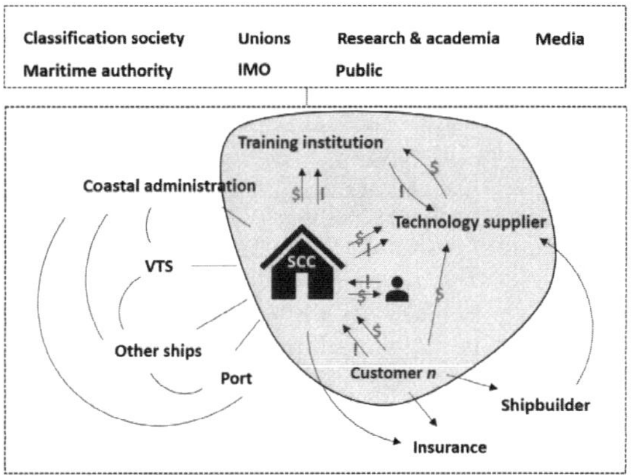

Figure 5: An example of a critical stakeholder analysis (Exclamation marks represent direction of complaints. The person icon represents an SCC operator.)

The understanding of stakeholder inter-relationships may also serve to frame the importance of addressing human factors issues encountered in the SCC working environment. Every stakeholder will be affected by poor management of the operators' human factors design needs and, conversely, can serve to remedy them. Potential remedies of human performance in automation is the subject of many publications [12]. Remedies for situation awareness, for example, may stem from adopting appropriate LOAs and even adaptive automation. Feedback to multiple sensory channels, like haptic or tactile displays, may counter multitask overload caused by over-used visual or audio channels. When designed correctly, automation in a system dependent upon human operators can be implemented in a way that increases margins of safety, results in the least possible disruptive manual takeover, and yields a satisfactory working environment.

6. CONCLUSIONS

This paper presented a stakeholder map of autonomous shipping centred around the SCC based on an aggregation of a large dataset collected from a workshop and multiple interviews. Understanding of the critical stakeholders was shown to be important because it can stymie potential design flaws. This is especially the case when addressing human factors issues. For the SCC, the human operator is expected to remotely monitor multiple autonomous ships and manually take over control in case of system failure. Three high-level human factors issues were identified in the findings (Section 4) and some remedies were proposed for design of effective human-automation systems (Section 5). The findings were also framed in an operational context using the stakeholder map and an example showing a possible exchange of complaints due to human factors issues was plotted onto the stakeholder map (Section 5).
Some important research questions may be summarized:

- How many vessels can one operator, or one team of operators, monitor simultaneously, depending on LOA, task similarity, and priority?
- How much time is required to take over control during an autonomous-to-manual handover, depending on LOA?
- What specific skills are needed? Are mariners with traditional seafaring backgrounds needed or can control room operators be trained according to the task requirements?

A technology-centred approach risks missing the human element that is needed to successfully supervise and operate highly automated systems. Focus on controls, software, and sensors will inevitably be of limited use if little attention is afforded to the human operators' needs in the larger system. The challenge, therefore, is to optimize human-automation performance with an HCD approach.

7. ACKNOWLEDGEMENTS

This research has partly been conducted with the HUMANE project (Human Autonomy Enable), which is funded by the Norwegian Research Council, whose contribution is hereby gratefully acknowledged.

8. REFERENCES

[1] M. Lutzhoft, A. Hynnekleiv, J. V. Earthy, and E. S. Petersen, "Human-centred maritime autonomy - An ethnography of the future," *J. Phys. Conf. Ser.*, vol. 1357, p. 012032, Oct. 2019.

[2] "IMO takes first steps to address autonomous ships." [Online]. Available: http://www.imo.org/en/MediaCentre/PressBriefings/Pages/08-MSC-99-MASS-scoping.aspx. [Accessed: 26-Nov-2019].

[3] "Auto-docking - a giant leap for merchant shipping," *Twentyfour7*. [Online]. Available: https://www.wartsila.com/twentyfour7/innovation/auto-docking-a-giant-leap-for-merchant-shipping. [Accessed: 09-Dec-2019].

[4] "Press releases." [Online]. Available: https://www.rolls-royce.com/media/press-releases.aspx. [Accessed: 09-Dec-2019].

[5] "Wilhelmsen and KONGSBERG establish world's first autonomous shipping company - KONGSBERG." [Online]. Available: https://www.kongsberg.com/news-and-media/news-archive/2018/wilhelmsen-and-kongsberg-establish-worlds-first-autonomous-shipping-company/. [Accessed: 09-Dec-2019].

[6] "Meld. St. 33: Nasjonal Transportplan 2018-2029." Det Kongelige Samferdselsdepartement.

[7] N. B. Sarter, D. D. Woods, and C. E. Billings, "Automation surprises," *Handb. Hum. Factors Ergon.*, vol. 2, pp. 1926–1943, 1997.

[8] M. Vagia, A. A. Transeth, and S. A. Fjerdingen, "A literature review on the levels of automation during the years. What are the different taxonomies that have been proposed?," *Appl. Ergon.*, vol. 53, pp. 190–202, Mar. 2016.

[9] Ø. Jan Rødseth, H. Nordahl, and Å. Hoem, "Characterization of Autonomy in Merchant Ships," in *2018 OCEANS - MTS/IEEE Kobe Techno-Oceans (OTO)*, 2018, pp. 1–7.

[10] On-Road Automated Driving (ORAD) committee, "Taxonomy and Definitions for Terms Related to On-Road Motor Vehicle Automated Driving Systems," SAE International.

[11] Ø. J. Rødseth, "Definitions for Autonomous Merchant Ships," p. 22.

[12] C. D. Wickens, J. G. Hollands, S. Banbury, and R. Parasuraman, *Engineering psychology and human performance*. Psychology Press, 2015.

[13] C. E. Billings, *Aviation automation: The search for a human-centered approach*. CRC Press, 2018.

[14] R. K. Yin, *Case study research and applications: Design and methods*. Sage publications, 2017.

[15] L. D. Loukopoulos, R. K. Dismukes, and I. Barshi, *The multitasking myth: Handling complexity in real-world operations*. Routledge, 2016.

[16] M. R. Endsley, "Design and evaluation for situation awareness enhancement," presented at the Proceedings of the Human Factors Society annual meeting, 1988, vol. 32, pp. 97–101.

[17] K. M. Donaldson, K. Ishii, and S. D. Sheppard, "Customer Value Chain Analysis," *Res. Eng. Des.*, vol. 16, no. 4, pp. 174–183, Apr. 2006.

[18] B. Buxton, *Sketching user experiences: getting the design right and the right design*. Morgan kaufmann, 2010.

[19] D. Norman, *The design of everyday things: Revised and expanded edition*. Basic books, 2013.

9. AUTHORS BIOGRAPHY

Erik Veitch is a PhD candidate at the Department of Design at the Norwegian University of Science and Technology (NTNU). He is responsible for applied research in the field of interaction design to test and develop solutions for the SCC. His previous experience includes simulator studies to measure human performance variability among seafarers tasked with demanding ship manoeuvres.

Agnieszka Hynnekleiv is a PhD candidate in Maritime Human Factors at Western Norway University of Applied Sciences and a team member of HUMANE project. Her doctoral research investigates future skills for the maritime personnel. She holds a master's degree in psychology and her previous experience includes designing training programmes for different occupational groups.

Margareta Lützhöft is Professor of Maritime Human Factors in the MarSafe group at the Western Norway University of Applied Sciences. Her research interests include Human Factors, human-centered design and the effects of new technology.

HUMAN FACTOR AND AUTOMATION: A STPA APPROACH TO ENHANCE SHIP SAFETY

P. Gualeni and **C. Bongermino,** University of Genoa, IT

SUMMARY

Ships complexity is continuously increasing and lots of action controls, traditionally carried out by crew members, are now managed and executed by the on-board automation systems. The relation among humans and automation system shall be studied with specific attention to their mutual interface and interaction issues, in order to better pursue safety. The STAMP (System-Theoretic Accident Model and Processes)-based tool named STPA (System Theoretic Process Analysis) consists the following steps: identify system hazards; draw functional control structure; identify unsafe control actions (UCAs); identify accident scenarios; formulate decisions and recommendations. In this paper an approach to model the relations among human operator and automation, based on STPA, will be proposed. An application case will be developed considering a large passenger ship and the specific hazard of dead ship condition (energy black-out). In case of navigation close to the shore and heavy weather condition, this situation can rapidly evolve into a ship loss.

1. INTRODUCTION

Cruise ships are among the most complex vessels on the seas, due to their operational profile, entailing thousands of people on board, and the customer oriented solutions installed.

In this context, the energy production and delivery system is a central asset and, during the design phase, it requires a strong perspective on the operational life of the ship. An important part of the power production is meant for the ship propulsion, nevertheless also the hotel loads are significant drivers for the design trade-off. In both cases, i.e. propulsion and hotel services as users, the proper power generation and distribution system is of vital importance in terms of safety and also commercial credibility.

Nevertheless, it is evident the increasing complexity of such energy systems during the latest decades and this, in turn, implies also an increasing reliance on automation control in order to better manage and coordinate the services of electrical power generation and delivery. It is recognized that automation can increase system efficiency and safety. However, it is not enough recognized that the role of human operators becomes even more important when automation becomes more significant and largely applied.

For the purpose of this paper, we will refer to the dead ship condition as the situation where the ship has just the emergency generator/s working (in compliance with IMO SOLAS Convention), but there is no power for ship propulsion and manoeuvring. The dead ship condition is an hazard state that, in case of adverse weather and proximity to the shore, can rapidly evolve into the ship loss.

The safety implications of power availability on board are so relevant that from June 2010 the Safe Return to Port [1] requirement has been introduced for passenger ships. The relevant outcomes of this rule, in terms of design features of modern large passenger ships, is an increased redundancy and segregation on board with focus on vital systems (specifically identified), among which propulsion, steering systems and electrical power delivery plant.

Nevertheless, it is well-known that safety is not only a matter of redundancy and systems availability; in fact, also for ships complying with the Safe Return to Port, blackout is still an issue worth to be investigated with a different perspective.

As already mentioned, automation is extensively exploited on board ships with the aim to provide higher efficiency and lower workload, assuming in some cases also a consequent reduction of human error probability. The point is that, beside redundancy, neither automation is sufficient to avoid human error [2, 3].

In general, more complex the systems are, more the final outcome depends also on the appropriate interface performance. This is the reason why it is important not just designing the automation with a focus on the physical technical aspects, like sensors and actuators, or the information aspects, like algorithms, but considering it as a joint human–automation system [4].

The joint approach is needed because automation implies a disruptive change in the feedback typology received by the human operator about the controlled physical system. This, in turn, influences the nature and the paradigm of the tasks carried out by the human operator. Automation requires new skills and, usually, it creates the shift of skills typology: from the simplest ones (training and/or ruled-based) to the knowledge-based ones. It is evident how education and training become more and more important.

"Low level" simple tasks can be carried out by the automation, therefore human operators are in charge of monitoring and intervene, only in case when "high level" control is required. But passive monitoring and supervisory control, without direct manual action on the system, are less involving and the operator might reduce the level of attention [5, 6]. The reduction or absence of direct action on the system, together with the reliance on automation feedback, can compromise the operator's mental model (discussed later in the paper), that might become inadequate to support correct decision-making about control of the system [7].

For a joint human–automation system analysis, a representation of the physical system, the human operator and the context is necessary [8].

In this paper, an approach to discuss the human-automation interaction for the management of the power generation and distribution system will be proposed and discussed. The application case will be developed considering the specific hazard of dead ship condition, i.e. of energy black-out on board.

To this aim, it seems very helpful the use of STAMP (System-Theoretic Accident Model and Processes) [9] methodology and, in particular, its safety control structure appears to be effective to model and reason about the best ways to enforce and implement safety from the top to the bottom of the structure.

2. THE SYSTEM THEORETIC PROCESS ANALYSIS (STPA)

System-Theoretic Accident Model and Processes (STAMP) is a top-down system-based accident model, that focuses on enforcing constraints rather than preventing failures, when pursuing safety performances. Safety is considered a dynamic control problem rather than a reliability problem. The system under investigation is described and modelled by a hierarchical safety control structure, where each part of the system is identified and its relationships with the other parts of the system are described. The main focus of this approach is to identify the safety constraints that are exerted from the safety control structure, because events leading to a loss can occur only when safety constraints from a higher level in the safety control structure are not enforced.

System-Theoretic Process Analysis (STPA) is the hazard technique built upon the foundation provided by STAMP, that mainly consists of collecting and structuring information about the system and its performances, identifying unsafe control actions and causal factors of unsafe actions.

In this perspective, STPA is applied as a very effective way to understand where and how performance variability might happen [10]. Therefore, it can also suggest how to better handle situations.

The roadmap of STPA consists of four steps (Figure 1): define the purpose of the analysis (identify losses and hazards, define system boundaries), model the control structure, identify the unsafe control actions, identify loss scenarios [11].

The first step in applying STPA is to define the purpose of the analysis and it mainly consists of identifying the losses to prevent. A *loss* can be defined as "something of value for the stakeholders" and it can include a loss of human life or human injury, property damage, environmental pollution, loss of mission, loss of reputation, loss or leak of sensitive information, or any other loss that is unacceptable to the stakeholders. Before starting the STPA analysis, the stakeholders must identify the losses on which they want to focus.

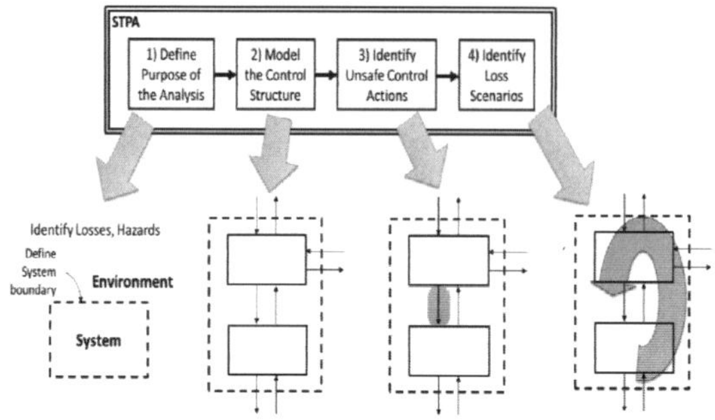

Figure 1: The four steps of STPA [11]

The next step is to define the hazards related to these losses. A *hazard* is a "system state or set of conditions" that, in combination with some other circumstances, will lead to a loss. To do that, it is needed, at first, to identify the system under investigation and its boundaries. The goal of engineering is to eliminate or mitigate the effects of hazards in the system under analysis, so some levels of control are necessary.

The second step consists of modelling the control structure, that is a hierarchical system model composed of feedback control loops. An effective control structure enforces constraints on the behavior of the overall system. A basic control structure is composed of control loops, as shown in Figure 2: a controller provides control actions to enforce constraints on the behavior of the controlled process.

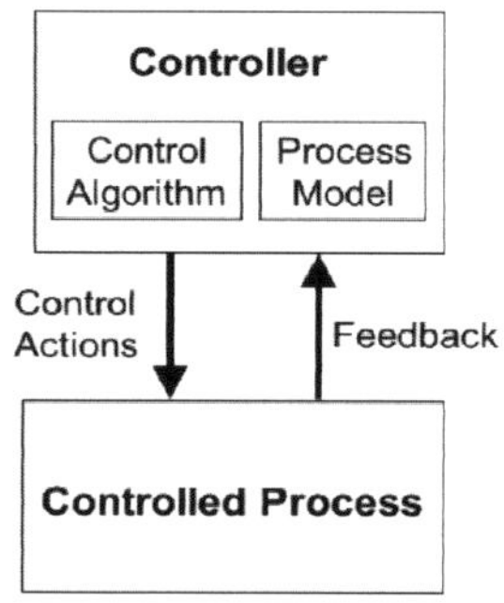

Figure 2: Basic elements of a control structure [11]

The controller is composed of the control algorithm and the process model. The control algorithm represents the controller's decision making process: it determines the control actions to provide. The process model represents the controller's internal beliefs used to make decisions. Moreover, process models may include beliefs about the process being controlled, or other relevant aspects of the system, or the environment. Process models may be updated in part by feedbacks used to observe the controlled process. In this perspective, the control structure can be used to explain and anticipate complex system behavior and critical issue.

The third step is about unsafe control actions (UCAs) identification. An UCA is a control action that, in a particular context and worst-case environment, will lead to a hazard. There are four ways a control action can be unsafe:

- not providing the control action leads to a hazard;
- providing the control action leads to a hazard;
- providing a potentially safe control action but too early, too late, or in the wrong order;
- the control action lasts too long or is stopped too soon (for continuous control actions, not discrete ones).

Typically, UCAs are shown in tables, as in Figure 3, with each of these four types in a separate column and each control action in a separate row.

Control Action	Unsafe Control Actions			
	Not Providing Causes Hazard	Providing Causes Hazard	Wrong Timing /Order	Duration Too Short/Long
A	Controller does not provide "A" when... *context(s)* *which lead to [Hazard(s)]*	Controller provides "A" when... *context(s)* *which lead to [Hazard(s)]*	Controller provides "A" too early / too late when... *context(s)* *which lead to [Hazard(s)]*	Controller provides "A" too long / too short when... *context(s)* *which lead to [Hazard(s)]*
B	...	...	...	...

Figure 3: The four types of unsafe control actions [12]

An UCA should specify the context in which the control action is unsafe. In this way, it is possible to eliminate those instances from the system design or find ways to mitigate them. In defining an UCA, environmental conditions, controlled process states, controller states, previous actions can be included. The hazards resulting as the outcome of the UCA are to be indicated.

The fourth step is to identify loss scenarios. A loss scenario describes the causal factors that can lead to the unsafe control actions and to hazards. Two types of loss scenarios must be considered, that correspond to the answers to these questions (Figure 4):

- why would unsafe control actions occur?
- why would control actions be improperly executed or not executed, leading to hazards?

The goal of these scenarios is to explain why, through a combination of factors, the unsafe control action may appear reasonable in the context taken into consideration. Causal scenarios often involve the concept of process models, the internal representations that a controller uses to understand the controlled process. This model must include relationships among system variables, the current state of those variables and ways in which the process can change state.

Factors that contribute to hazardous states may occur anywhere in the control loop. Possible control loop flaws can be inadequate sensor operation, missing feedback and inconsistent process models. Flaws in one zone of the control loop could easily propagate to other parts of the control loop: for example, a measurement inaccuracy could lead to operators receiving inadequate feedback, leaving them with an incomplete model of the controlled process.

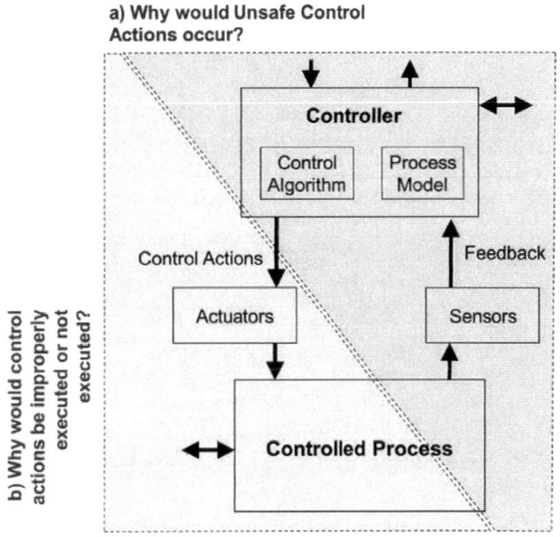

Figure 4: Identification of Loss Scenarios [11]

It is important to note that the STPA model is not intended to serve as a checklist and single causal factors in isolation do not explain unsafe control actions. To write meaningful scenarios requires addressing why the system as a whole can experience dysfunctional interactions that lead to hazardous states.

2.1 STPA MODEL FOR HUMAN CONTROLLERS

A control loop inclusive of a human controller is depicted in Figure 5 [12]. It is worth mentioning the introduction of a (mental) model of the automation in addition to a (mental) model of the controlled process is necessary.

From this model, it may be considered that the human controller could have an incomplete or incorrect model of the automation, but it is up to the analyst to determine exactly what types of flaws are implied and to identify any related accident causal scenarios.

In order to create the causal scenarios related to human controllers, it is necessary to further detail the relevant model as evidenced in Figure 6 [12].

The three parts (from left to right: Control Action Selection, Mental Models and Mental Model Update) correspond to three important questions to be answered while writing scenarios related to human control actions:

- how did the operator choose which control action to perform?
- what does the operator know or believe about the system?
- how did the operator come to have their current knowledge or beliefs?

In the following, each of the three points is further examined.

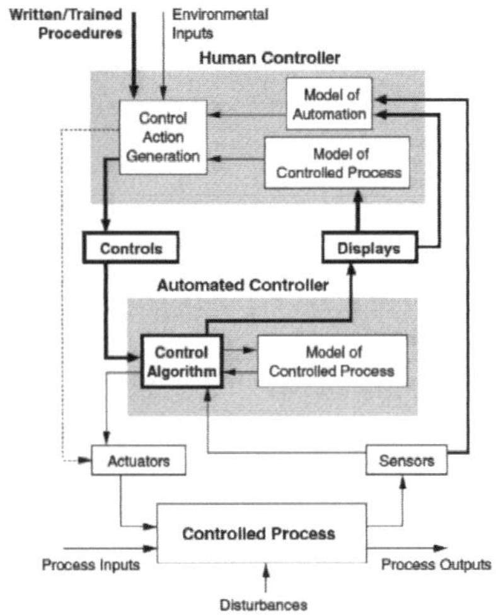

Figure 5: control loop including a human controller [12]

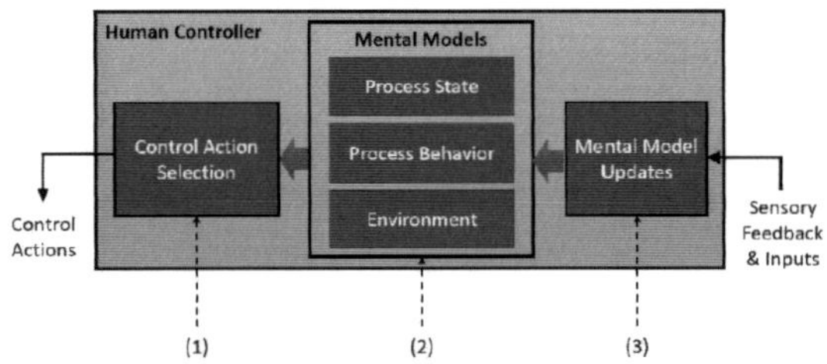

Figure 6: Model for the Human Controller [12]

Control Action Selection: *how did the operator choose which control action to perform?*

The Control Action Selection phase corresponds to the Control Algorithm in the automation controller model and it aims to explain why a particular control action is chosen. Unlike software decisions, the way humans make decisions might depend on the context. Several factors that affect the way the human operators decide can be identified.

Firstly, it may be relevant to consider the operator's goals. The goals of the operator can differ from the goal of the system designer, so this divergence in goals may imply that the operator takes actions that the system designer would not have predicted and analyzed.

Moreover, decisions can be very different in nature: some of them are made more rapidly, or automatically, than others. Rasmussen [13] defines these as "skill-based" decisions, that simply require recognizing a familiar situation and performing the known action. Other decisions can be made by simply categorizing a situation and applying appropriate rules ("rule-based"). However, most unusual situations that the human controller may experience require "knowledge-based" decision making. This requires mental simulation based on what it is known about the system, in order to identify a solution that the human operator assumes will be appropriate. Considering whether a decision is skill-based, rule-based or knowledge-based can help, when using STPA, to evaluate why such decision has been selected.

Usually, automation system can easily take care of and implement tasks that for human controller are skill-based or rule-based, but the knowledge-based ones are very likely to remain in charge of the human controller.

Decisions taken relying on the combination of knowledge and suitable mental models (based, for example, on previous experience) are supposed to be successful; however, some contingent situations (environmental o psychological) can compromise the selection of the right choice [14].

Mental Model: *what does the operator know or believe about the system?*

In literature, there are several definitions of mental models across different disciplines, but, in the context of STPA, mental models can be understood as cognitive representations of the world, i.e. operator's opinions and understanding about the current system condition. This mental models can be used to anticipate situation evolution and carry out trade-off assessments [15]. However, they might be incomplete or not perfectly adequate [16] and this is crucial when discussing the human controller control action selection during a STPA analysis.

Mental Model Updates: *how did the operator come to have their current knowledge or beliefs?*

Mental models are very important and they need to be continuously completed and updated.

The most important source for this feeding process is the interaction with the physical system itself and the automation system, able to provide updates for the respective mental models. To this aim, usually the feedback channel provides the flow of input for the mental model update. Nevertheless, it is worth mentioning that excessive source of data can be even detrimental.

The methodology described above appears to be suitable in order to investigate the interactions between the automation and the human operators, when dealing with the management and control of the power generation and distribution system on board cruise ships. An application case will be developed in order to discuss some critical issues that can be, in principle, overcome by appropriate design solutions, technical specifications and personnel training.

3. APPLICATION CASE

3.1 DESCRIPTION OF THE ENERGY SYSTEM

In the investigated case, four gen-sets are installed, able to supply final users by means of two main switchboards. The generation and distribution system is intrinsically redundant. In this case, the propulsion is delivered by two electrical engines.
An efficient use of the power system relies on the automation support in order to better follow the real electrical power needed and to provide the proper amount of power supply.
Visualising the interaction among the human controller, the automation system and the power system is, therefore, of outstanding importance to guide the design for increased ship safety and efficiency.

3.2 IDENTIFICATION OF THE STUDY

As a first step, losses that the analysis is aimed to prevent are to be defined. In Figure 7 a list is given.

LOSSES	
L 1	loss of human life
L 2	human injury
L 3	enviromental pollution
L 4	property damage
L 5	loss of reputation

Figure 7: Losses identification

In Figure 8 the specific hazards that are in relation with the blackout condition are reported.

HAZARD	
H 1	the ship is without propulsion
H 2	the ship is without hotel services
...	...

Figure 8: Hazard identification

3.3 SAFETY CONTROL STRUCTURE

The kernel of STPA is the safety control structure. As already explained, it is a hierarchical model of the system, where feedback and control actions can be represented and therefore discussed. For the purpose of this study a safety control structure has been developed and it is represented in Figure 9.
The human operator model presented in 2.1 is implemented. The level of details captured by the safety control structure is what it is considered necessary to develop the application case. For example, the human operators might be further split evidencing the different jobs and tasks of each of them. A division might be described among human operators working in the engine control room and the ones on site close to the physical system.

In Figure 10 the frame of the specific part of the safety control structure that will be investigated has been put in evidence. In the diagram, the control actions and the feedbacks useful for the investigation are expressed.
More specifically, the application case will focus on the gen-sets and the distribution system (since they are physically at the base of the blackout issue on board) and their interaction with the automation and the human operators as controllers, in a hierarchical perspective.

3.4 UNSAFE CONTROL ACTION

On the base of the safety control structure, then unsafe control actions can be identified.
As already defined, an unsafe control action (UCA) is a control action that, in a particular context and worst-case environment, will lead to a hazard. If the control structure is well developed, the identification of the UCAs is a systematic activity and it is driven by the syntax rule composed of the following five parts:
<Source> <Type> <Control Action> <Context> <Link to Hazards>.
With reference to the specific area framed in Figure 10, more than one hundred UCAs have been identified and, for the purpose of the paper, the attention is focussed on the following:
«Automation» provides turn-off the «requested gen-sets» during navigation and there are not any other gen-sets to turn-on in the sequence [H1, H2].
In order to better identify how the hazards H1 and H2 can evolve in one of the considered losses, the above formulated UCA has been further detailed as:
«Automation» provides turn-off the «requested gen-sets» during navigation <u>in heavy weather conditions close to another vessel</u> and there are not any other gen-sets to turn-on in the sequence [H1, H2].

3.5 LOSS SCENARIOS

The further and last step is to identify the loss scenarios related to the above mentioned UCA:
«Automation» provides turn-off the «requested gen-sets» during navigation in heavy weather conditions close to another vessel and there are not any other gen-sets to turn-on in the sequence [H1, H2].
This step is developed answering to the two types of question previously presented in this paper:
α) why would unsafe control actions occur?
β) why would control actions be improperly executed or not executed, leading to hazards?
These questions, in turn, imply considerations about:
α1) unsafe controller behaviour;
α2) causes of inadequate feedback /information;
β1) control path;
β2) other factors related to controlled process.
After a general overview on all of them, further insight has been developed for α1).
There are four general reasons that can cause an unsafe controller behaviour (α1) [11]:

- failures involving the controller;
- inadequate control algorithm;
- unsafe control input (from another controller);
- inadequate process model.

At the same time, looking at the control structure it is self-evident that over the automation, hierarchically, there is the human controller and the model is able to evidence the interface between the human factor and the automation system. For example, the automation system is a controller that, in turn, is made of a physical system that can suffer a failure. In such case, the human operator is called to directly intervene as back-up controller.

Getting back to the loss scenarios identification, in this application case, the issue *unsafe control input by another controller* is further investigated. Thanks to the safety control structure, it is possible to analyse the effect of an *unsafe control input by the human controller,* immediately above the automation controller in the model. The relevant loss scenario is, therefore, formulated as follows:

The human controller acknowledged an alarm related to the gen-sets status without any follow-up action. The «automation behaviour» turns-off the requested gen-sets. There aren't any available gen-sets to turn-on in the sequence to deliver the requested power and, as a result, the ship suffers a blackout [H1, H2].

«Automation behaviour» means the actions of the automation system as consequence of its control algorithm and process model. In this case, the «automation behaviour» consists of the turn off the gen-set, when the functioning parameters are out of the suitable range of safeguard for the gen-set itself.

Other loss scenarios can be formulated from the identified UCA. However, for the purpose of this paper the analysis is focused only on the above identified one. The selected UCA is very powerful to put in evidence the interaction between the human operators and the automation, when discussing about possible ship blackout hazard. In the investigated scenario, the undesirable blackout is produced by automation, in turn originated by a bad interaction with human controller at an upper level of safety control structure. It is interesting to discuss why a clumsy interaction between the human operator and the automation can occur, i.e why the human controller doesn't perform any follow up after the acknowledgement of an automation alarm.

In this perspective, for each of the three parts of the human controller model discussed above (control action selection, mental model and mental model updates) some questions have been selected [12].

1) Control Action Selection: How did the operator choose which control action to perform?
   - What were the operator's goals? How might they differ from the goals of the system designer? Does the operator have multiple, or conflicting goals?
   - How much experience did the operator have with the system or other similar systems? Was the situation simple and familiar, or novel and complex?
   - Was the operator in a complex or novel situation that required *knowledge-based* selection? How did their mental models contribute to the decision they made?

2) Mental Model: *What does the operator know or believe about the system?*
   - What triggers mode changes in the system?
   - What actions are available to the operator in the current mode?
   - What is the relationship between operator inputs and system behavior?
   - What does the operator believe other controllers in the environment will do?

3) Mental Model Updates: how did the operator come to have their current knowledge or beliefs?
   - How much effort is required for the operator to access necessary information, including process states?
   - What does the operator believe is the most important to monitor?

Considering the questions above, further details can be added to better understand the scenario also in terms of causal factors. A wider description of the scenario can be formulated bearing in mind what follows.

The human controller decided that it was not necessary to apply a control action in response to the automation alarm. The decision can be based on an inadequate mental model, that creates the following effects: underestimation of the alarm importance or non-reliance on the automation detection.

In particular, the decision can be a consequence of a prioritization of urgent actions regarding other several tasks to be executed by the operator at the same time. Moreover, previous experiences can suggest that the alarm generated by the automation is of low priority, even though the actual context, not familiar to the operator, can have an influence on the matter and can imply an upgrading of the urgency.

With regard to the non-reliance on the automation system, it is possible that the operator assumes that the external conditions (large roll motions and accelerations) are having an influence on the automation sensors reliability and that alarm on gen-sets can be ignored. The operator can believe for example that in other external conditions the automation wouldn't raise such alarm.

3.6 STPA OUTCOMES

In principle, from an STPA analysis it is possible to derive several outcomes to improve the system safety [11].
For the topic investigated in this paper the results of STPA can consist of:

- driving the system architecture;
- creating executable requirements;
- identifying design recommendations;
- identifying mitigations and safeguards needed;
- defining test cases and create test plans;
- developing leading indicators of risk;
- designing more effective safety management systems.

The human controller in the control structure plays the role of the controller of the automation system, so he has also the task to detect when the automation is going to adopt an unsafe behaviour.
Especially when safety implications are expected, an enhanced level of interaction between human operator and automation should be imagined and implemented. For example, in specific cases like engine shut down, the operators should be "consulted" by the automation system for the final executive authorization.
In this perspective, the training and education of the human controller is the key for improving the whole system safety, especially, when knowledge-based control actions have to be thought and executed.
The training and education program should focus on enhancing the experience of the operator with the automation system, for example, engaged in complex or novel operational conditions.
In particular, the training could enhance the capability to pay a more thorough attention to the power system conditions and efficiency in adverse weather conditions.
At the same time, during design technical specifications all the possible operational scenarios shall be identified in order to define the operational domain of on board systems (for example with reference to roll angle and/or accelerations in heavy seas). In particular, sensors, the automation system relies on for raising alarms, should result able to function properly in the defined operational scenarios.

4. CONCLUSIONS

In the paper the possibility to apply STPA technique to analyse the relation between human operators and automation has been investigated.
An application case has been carried out with reference to a large cruise vessel. The specific issue of black-out on board has been selected and the hazard situations with the ship without propulsion and/or hotel services have been identified.
Starting from the safety control structure, a particular unsafe control action has been selected for the development of a specific loss scenario, significant to put in evidence the relevance of the interaction between human controllers and the automation system.
STPA has enabled the visualization of the hierarchy among the ship energy system, the automation system and the human operators. This is useful to discuss in a design stage the characteristics and the logic of the automation system (integration with crew members in decision making included), especially when some disruptive conditions, like heavy weather, can characterize the scenario and a more detailed mental model is needed for the proper decision making.
The loss scenarios identified using STPA can be the track where to root training and education subjects, in order to enhance the operators' capability of making decisions in complex or novel operational situations.
Moreover, in a design for operations perspectives, it has been put in evidence how the design of automation, with reference to technical aspects of sensors, algorithms and actuators, should be developed with the aim to deliver reliable data, also in complex or novel situations and in a joint human–automation approach.

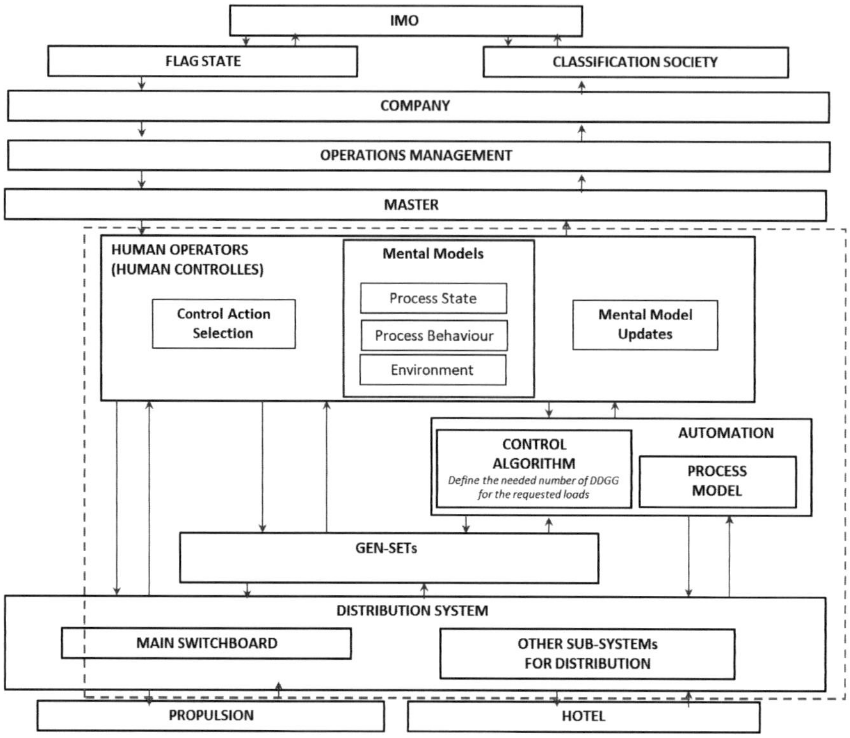

Figure 9: The safety control structure

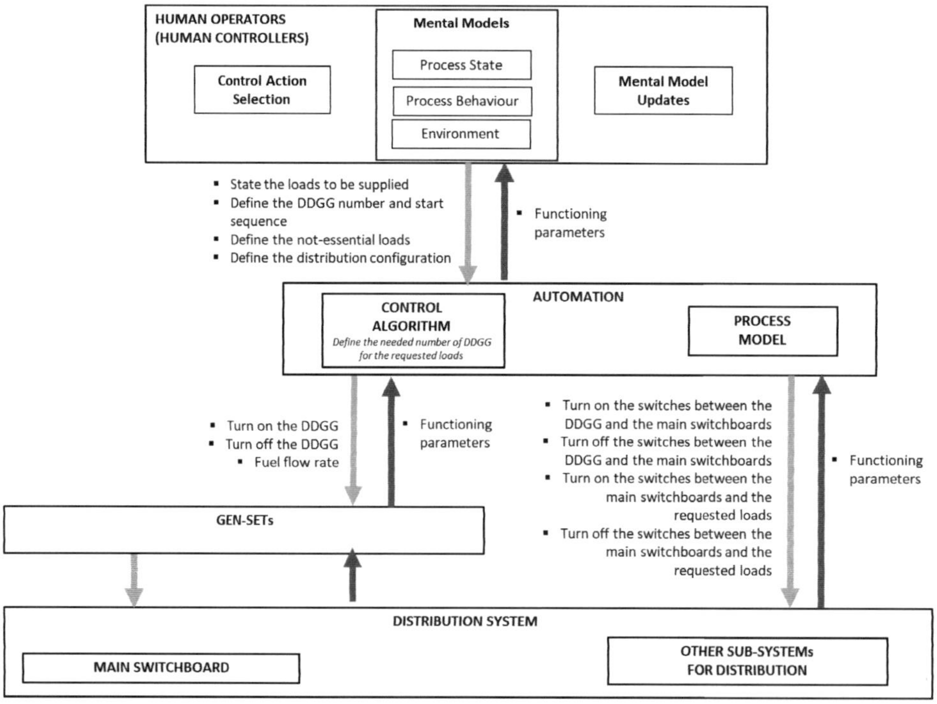

Figure 10: Focus on the investigated part of safety control structure

5. REFERENCES

1. IMO, 'Resolution MSC.216(82) Amendments to the International Convention for the Safety of Life at Sea, 1974, as amended', 2006.
2. NTSB 'Marine accident report – Grounding of the Panamanian Passenger Ship ROYAL MAJESTY on Rose and Crown Shoal Near Nantucket, Massachusetts' June 10, 1995.
3. LUTZHOFT, M. H., DEKKER, S.W.A., 'On your watch: Automation on the bridge', *Journal of Navigation* 55(1), 83–96, 2002.
4. LEE, D. J., SEPPELT, B. D., 'Human Factors in Automation Design', *Handbook of Automation*, chapter 25, 2009.
5. SHERIDAN, T.B., 'Supervisory control', *Handbook of Human Factors*, ed. by G. Salvendy , 1987.
6. EPHRATH, A.R., YOUNG, L.R., 'Monitoring vs. man in the loop detection of aircraft control failures', *Human Detection and Diagnosis of System Failures*, ed. by Rasmussen J., Rouse W.B., 1981.
7. MORAY, N., 'Human factors in process control', *Handbook of Human Factors and Ergonomics*, ed. by Salvendy G., 1997.
8. KIRLIK, A., MILLER, R.A. , JAGACINSKY, R.J., 'Supervisory control in a dynamic and uncertain environment: a process model of skilled human-environment interaction', *IEEE Transactions on Systems, Man, and Cybernetics*, 1993.
9. LEVESON, N.G., 'Engineering a safer world: Systems thinking applied to safety', *MIT Press*, 2011.
10. LEVESON, N.G., DULAC, N., ZIPKIN, D., CUTCHER-GERSHENFED, J., CARROLL, J., BARRETT B., 'Engineering Resilience into Safety-critical Systems', *MIT Press*, 2006.
11. LEVESON, N.G., THOMAS, J.P., 'STPA Handbook', 2018.
12. FRANCE, M. E., 'Engineering for Humans: A New Extension to STPA', *Thesis of MSc in Aeronautics and Astronautics at MIT*, 2017.
13. RASMUSSEN, J., 'Skills, Rules, and Knowledge; Signals, Signs, and Symbols, and Other Distinctions in Human Performance Models', *IEEE Transactions on Systems, Man and Cybernetics*, Vol. SMC-13, No. 3, 1983.
14. DEKKER, S. W. A., 'The Field Guide to Understanding Human Error', *1st ed. Ashgate Publishing, Ltd.*, 2006.
15. CRAIK, K., 'The Nature of Explanation', *Cambridge University Press*, 1943.
16. JOHNSON-LAIRD, P. N., 'The History of Mental Models', *Psychology of Reasoning: Theoretical and Historical Perspectives*, ed. by Manktelow K. and Chung M. C. New York Psychology Press, 2004.

6. AUTHORS BIOGRAPHY

Paola Gualeni is graduated in Naval Architecture and Marine Engineering at the University of Genoa where she holds the position of Associate Professor. She teaches Ship Hydrostatics and Stability and her main fields of interest are Ship Safety and Innovative Ship Design Methodologies.

Costantino Bongermino in 2017 graduated in Naval Architecture and Marine Engineering at the University of Genoa where at present he holds the position of Ph. D fellow with specific interest on human factor integration in maritime systems safety, from both the design and the operations perspective. He serves as Sub-Liutenant in Italian Navy on the Italian aircraft Cavour.

DEVELOPMENT, TEST AND SELECTION OF A SHOCK MITIGATION SEAT FOR AN ALL-WEATHER LIFEBOAT

T. Gunston, CDG Coast Dynamics Group, Canada
A. Clarke, Royal National Lifeboat Institution, UK

SUMMARY

Shock and vibration exposures in fast boats at sea can be more severe than in any other mode of transport. Shock mitigation seating plays a key part in reducing the possibility of injury, but seats must be suitable for the boat and for the way the boat will be used.

The RNLI recently conducted an innovative sea trial programme to assess the performance of eight seats, from three manufacturers, for use in the Severn Class all-weather lifeboat. The trials were designed to maximise data capture while minimising crew exposure. Seat performance was measured with a range of occupants and operating conditions and the final seat selection was based on analysis of over 2,500 wave impacts. These "wave by wave" trials involved normal boat operation and were much more efficient to carry out than the more conventional, but often impractical, approach of attempting to carry out controlled, timed runs at specific speeds in specific sea states.

Ahead of the trial, the seat manufacturer used information on RNLI crew demographics and craft operating conditions to optimise the seat. The seat configuration was determined from laboratory drop-tests at their in-house facility, with parameters chosen that were most likely to perform well in the Severn. The laboratory test procedure followed the methods defined by the UK's Ministry of Defence Protocol 1&2 and the US Navy's NSW/CCD-080-TR-2015/010 seat testing requirement.

This paper describes the sea trials approach and seat design optimisation that led to the selection of the seat for use in the Severn Class.

1. INTRODUCTION

Shock and vibration exposures in fast boats at sea can, in some operating conditions, be very severe. The motion is often perceived as a sequence of impacts rather than a vibration. A deck acceleration recorded on a planing or semi-planing craft heading into an oncoming sea may be described as a series of wave impact shocks overlaid on an underlying vibration, as shown in Figure 1.

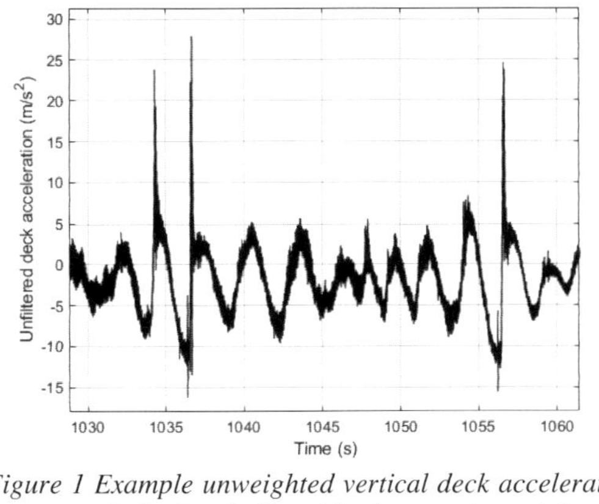

Figure 1 Example unweighted vertical deck acceleration on a semi-planing hull, with three substantial wave impact events

Reducing the risk associated with such exposures requires a holistic approach. Fitting suspension seats is one part of such an approach, but assessment of the suitability of such seats presents a number of challenges.

The first stage of seat selection should be to ensure it is generally suitable for the type, size, power and intended use of the boat. Laboratory tests exist for many suspension seats, including those intended for use in earthmovers [1], agricultural tractors [2], and the recently published US Navy requirement for marine craft [3,4]. These tests are intended to show that, in controlled conditions, a seat is likely to be reasonably suitable for its intended use. An inappropriate seat can make exposures substantially worse as some of the results below illustrate. Controlled laboratory conditions are necessarily much simpler than the real situation, however, so it is good practice to make final selection using more realistic trials.

Suspension seats are generally passive vibration isolators and so respond differently to different frequencies of motion. They are also extremely non-linear with input magnitude, from friction locked at low magnitudes up to the occurrence of suspension overtravel events at high magnitudes, involving impacts with travel stops and possibly loss of contact between the occupant and the seat. Figure 2 and Figure 3, from Wu and Griffin [5] and Gunston [6] illustrate some of these behaviours.

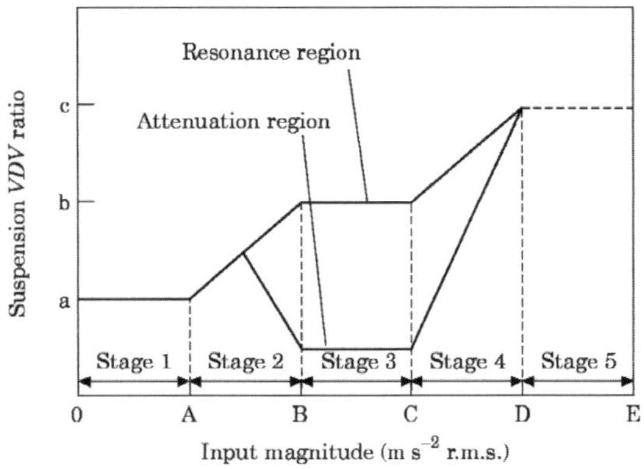

Figure 2 Characterising suspension seat behaviour with magnitude, summarised from Wu and Griffin [5], the attenuation region refers to response to motions consisting predominantly of frequencies sufficiently high for the suspension to attenuate when moving freely, the resonance region refers to motions with frequencies predominantly in the vicinity of the primary resonance frequency of the seat-person system. Stages are 1: Low magnitudes, friction locked; 2: Breaking away from friction; 3: Moving freely within the available stroke; 4: End-stop impacts / overtravel; and 5: Severe end-stop impacts

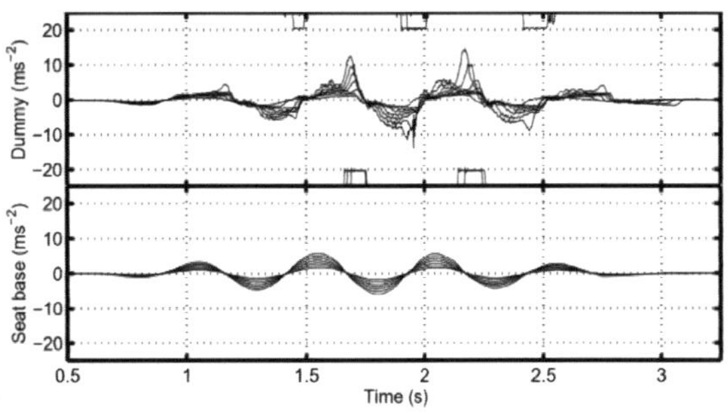

Figure 3 Example response of a suspension seat to seven increasing magnitudes of a windowed sinusoidal motion. Also shown are the signals from microswitches detecting seat end-stop impacts / overtravel events. This example is from Gunston [6] and uses an earthmover suspension seat tested in the laboratory with a passive anthropodynamic dummy as described by Lewis and Griffin [7].

Assessment of suspension seat performance in a land vehicle may be carried out by running a vehicle over various test surfaces at various speeds and with various seat occupants and then calculating the seat performance for each test run. Standards exist describing this process, for instance ISO 5008 [8] for agricultural tractor seats.

A similar process can be, and has been, used for sea trials. A trial to compare suspension seats typically requires repeated, comparable tests and there is often space for only two or three seats mounted on the same boat side by side at any time. It can be extremely time consuming to obtain reasonably well-matched conditions for sustained periods over several days.

The test team must typically commit to operating the boat on a particular day, so are at the mercy of the weather. The sea conditions could be milder, or much worse, than the trial requires. The motion experienced on different headings with respect to the prevailing are likely to be different so several test runs may be needed to assess each seat configuration. Even on days when conditions are appropriate, sea conditions will vary throughout the day with changes in wind and tide. Sea conditions might be reasonably consistent for short time periods in a given location, but the heights and shapes of consecutive waves will not be identical. Two short duration repeat tests conducted one after the other in the same direction over the same piece of water will not be identical.

The seat occupant will also introduce variations in response including intra-subject variability in the impedance of the body [9] and from passive or active reactions to anticipated inputs, for instance by bracing legs or tensing up before a big wave.

A sea trial of a suspension seat is therefore an attempt to quantify the performance of a highly non-linear mechanical system occupied by an actively varying impedance in response to an input motion that cannot be accurately controlled.

In this study, rather than carry out a planned series of test runs in specific conditions, we instead considered the seat response to each wave impact separately. Using this approach it is possible to assess seat response to mild, moderate and severe wave impacts irrespective of when they occurred during a sea trial.

This approach is intended to obtain as much seat performance data as possible while at the same time making trials more tolerant of changing or inconsistent sea conditions. Both factors help to keep crew vibration exposures as low as possible.

2. THE "WAVE-BY-WAVE" METHOD

This wave-by-wave method is illustrated using a series of trials conducted in 2018 to compare eight seats under consideration for use in an offshore lifeboat.

The method involves the following steps:

- measurement of vertical acceleration on, and below, seats in relevant, but not necessarily consistent, sea conditions;
- data quality check;
- frequency weighting;

- identification and extraction of 10-second time history samples centred on each impact;
- calculation of Vibration Dose Values (VDVs) for each 10 second sample;
- calculation of seat performance ratios, or SEAT (pronounced 'see-at') values;
- grouping SEAT values by seat, occupant weight and wave impact severity; and
- statistical comparison of suspension seat performance.

2.1 ACCELERATION MEASUREMENT

Vertical acceleration time histories were recorded at the deck and on the seat surface for an all-weather lifeboat in various operating conditions from sea state 1 to 5 (using the Douglas sea state scale) on seven days between September and December 2018. Seats were occupied by various seat occupants from 65 to 114 kg standing weight. Eight seat variants were assessed, with two or three variants fitted approximately side-by-side at any one time.

Vertical accelerations at the deck and on each seat were measured using 3 single axis, and 3 tri-axial (only logging x and z channels) accelerometers. Seat surface measurements were made using seat pads as described in ISO 10326-1 [10], although manufactured of a slightly softer material than that specified, and deck accelerometers were positioned directly under each seat. Seat displacement and boat gyration data was also collected but was not used in this analysis. Accelerations were acquired at 5,000 samples per second to a laptop via a Fylde *Microanalog* data acquisition system. A typical deck and seat accelerometer layout is shown in Figure 4.

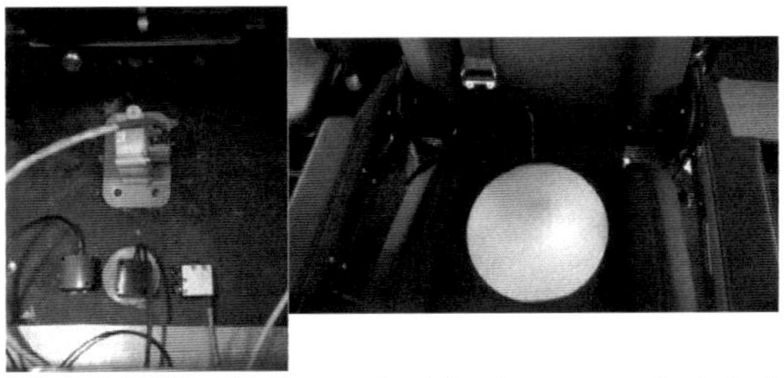

Figure 4 The picture to the left shows a typical deck sensor layout with a 2 tri-axial accelerometer (bottom left), rate gyro (bottom right), and linear string potentiometer (centre back). The image to the right shows the seat pad with the single axis accelerometer mounted inside.

2.2 INSPECTION AND WEIGHTING

The acceleration time histories were inspected for measurement artefacts or other data quality issues. No substantial issues were identified.

The time histories were then frequency weighted using the 'W_k' weighting as defined in ISO 2631-1:1997 [11].

This frequency weighting is used in the calculation of the Vibration Dose Value [12,13], and is also used in the calculation of the A(8) method referenced in the 2005 and 2007 UK Regulations on Control of Vibration at Work [14,15].

This weighting is applied to the complete time history, before any consideration of individual impacts. This was considered preferable to applying the weighting to truncated parts of the time history likely to be of similar, or shorter, duration than the impulse response of the W_k weighting filter function.

Other analysis methods exist, and some of these are discussed in a previous publication [16]. For the present analysis, the decision was taken to use W_k as defined in ISO 2631-1, rather than an alternative analysis method, because using the VDV is consistent with current risk assessment methods and associated Regulations as used in the UK; and we have seen no evidence that any other technique is a more appropriate predictor of injury risk.

2.3 IDENTIFICATION OF IMPACTS AND CALCULATION OF VIBRATION DOSE VALUES

Peak accelerations were defined as the maximum (upward) acceleration between two consecutive zero crossings of the normalised, W_k weighted acceleration time history as recorded on the deck, at the base of the seat.

Peak weighted accelerations below an arbitrary threshold of 0.5g were disregarded as they probably did not indicate a wave impact. A 10-second duration sample of the W_k weighted acceleration time history on the deck and on the seat was extracted, centred on each remaining peak as detected on the deck. By centring the sample on the peak of interest, a 2.5 second, cosine taper could be applied to more greatly attenuate any adjacent impacts without affecting the impact of interest. An example is shown in Figure 6. Again, the VDV of each tapered 10-second sample was calculated.

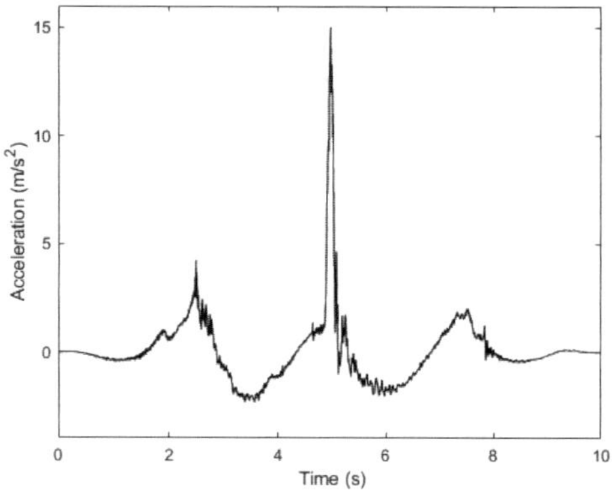

Figure 6 Example of a W_k-weighted, tapered, 10-second acceleration time history centred on an impact

2.4 CALCULATION OF SEAT VALUES, GROUPING AND AVERAGING

The ratio of the VDV recorded on the seat to that on the deck was calculated to obtain the seat performance ratio, or SEAT (pronounced see-at) value, for each 10-second sample centred on each identified wave impact. A total of 2,689 SEAT values were obtained, each nominally associated with the effect of one wave on:

a) the seat type and configuration;

b) the seat occupant weight; and

c) the wave severity, as defined by the VDV measured on the deck.

The results were grouped as follows:

- The range of seat occupant weights was divided into three weight classes, less than 80 kg overall standing mass, 80 to 100 kg, and greater than 100kg. These were not based on any anthropometric standard but were considered reasonably representative of the range of weights of fully dressed and equipped lifeboat crews.
- The range of wave severities was divided into 14 severity classes, a number arrived at empirically. Too few would provide too little resolution for assessing seat performance, but too many would result in many bins with few or no samples.
- The 2,689 SEAT values were therefore collated into 336 groups (14 amplitudes x 3 occupant weights x 8 seats) and the mean SEAT value for each of these 336 groups was calculated.

For the present analysis, no minimum number of samples in each of the 336 groups was specified for each average. Introduction of such a minimum would increase the confidence in each average but at the expense of discarding some results or reducing the resolution of the results. For the present analysis, a decision was taken to view all results at the reasonably high (14 step) wave severity resolution chosen, but to view the results in conjunction with the underlying values from which the average SEAT values were calculated.

3. RESULTS

The results are shown in Figure 7, with the deck motion for scaled in terms of 1-hour equivalent VDVs.

Trends similar to those in Figure 2 are apparent. In general, for motions with 1-hour equivalent VDVs of less than perhaps 20 $m/s^{1/75}$, all seats, on average, do not show substantial attenuation or amplification, although there is a lot of variation in individual results. As the amplitude increases some seats, on average, improve in performance while some get worse.

Feedback from trials crews suggested that the large amplification of Seat A with heavy subjects was associated with suspension overtravel, whereas seats that showed overall improvements with increasing amplitude were apparently freeing up from friction but remaining within their operational stroke.

4. SEAT COMPARISON

Overall seat performance was assessed using a series of two-sided non-parametric rank sum (Mann-Whitney U) tests on all SEAT values for all test conditions where the 1-hour-equivalent deck VDV exceeded 20 $m/s^{1.75}$. This threshold was used as a simple method of limiting the analysis to the more severe motions where good seat performance is more important.

This statistical test is essentially looking to determine which of each pair of seats shows the greatest median attenuation over all conditions in which it was measured, irrespective of the number of results obtained with different occupant weights or at different magnitudes.

Due to the variable quantity of data available for each seat for difference amplitudes and payloads, the use of parametric methods, or of more complex non-parametric methods, was not considered appropriate or useful.

Each seat variant was compared with each other seat with the significances obtained divided by the number of paired comparisons made.

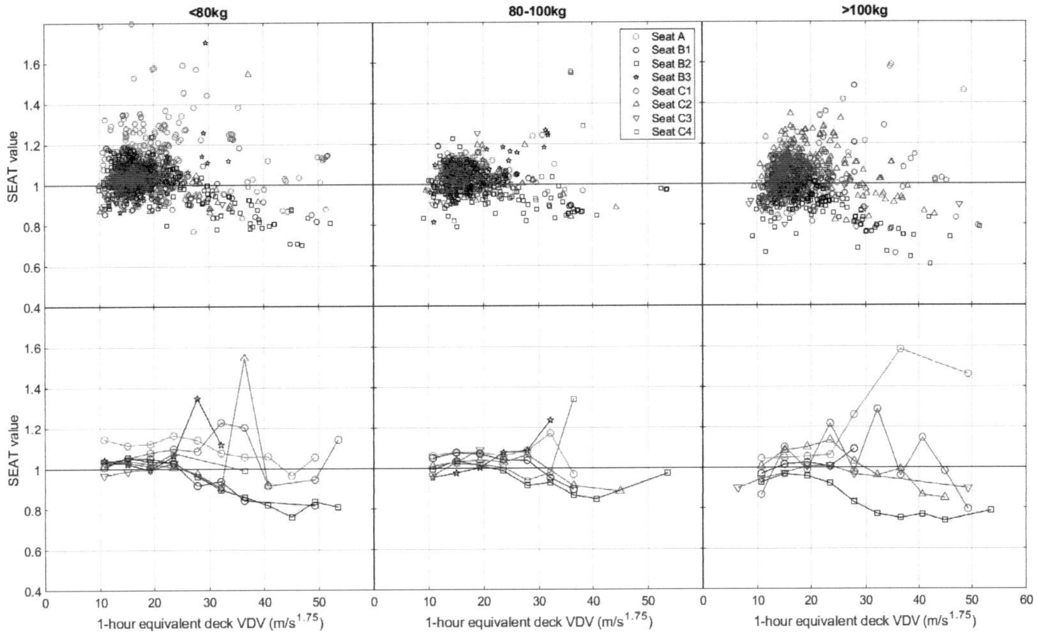

Figure 7 Individual (upper graphs) and average (lower graphs) SEAT values obtained with eights seat variants, various weights of test subject, in response to 10-second tapered samples of various magnitude of deck motion

This analysis, and inspection of the SEAT values obtained, suggested that the seats might be grouped into three performance classes.

- Group 1: Seat B2 only. This seat was significantly better (corrected Mann-Whitney U $p<0.05$) than all the others. The overall median SEAT value for the conditions assessed in the statistical analysis was around 0.9, indicating useful attenuation overall. As can be seen in Figure 7, however, the seat performance was substantially and consistently better than this for more severe conditions. Individual SEAT values for similar occupant weights and magnitudes were also less scattered compared to results obtained for some of the other seats, suggesting more consistent performance.

- Group 2: Seats B3, C2, C3 and C4. These seats were not significantly different from each other, although this was partly due to the large scatter in the SEAT values with some seats, and the limited dataset for some seats. Median SEAT values were around 1.0. This is probably better than could be achieved with a cushion seat, and greater attenuation was provided in some conditions.

- Group 3: Seats A, B1, and C1. These seats and generally did not perform well, and/or behaved inconsistently.

5. SEAT DEVELOPMENT

Seat B2 was a high performance marine sector specific suspension seat. Its suspension travel was extended for this application to accommodate the broadest user weight range and its shock-mitigating characteristics were optimized through a combination of simulations and drop-test experiments.

As an example of the testing carried out, Figure 8 below depicts an acceleration measured on the manufacturer's drop-test apparatus. This test was part of a series of guided vertical drops as defined by the marine seat test procedure [4].

The seat was fixed to a drop test platform with a 100 kg mass composed of steel-shot bags fastened to the seat pan. The deck was then raised and dropped from varying heights, with accelerometers placed at the seat base, on the seat pan, and on the cushion beneath the test mass.

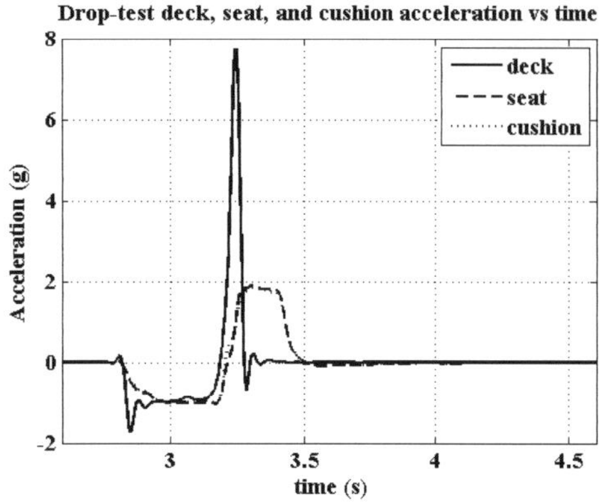

Figure 8. Accelerations measured on a sample seat drop-test. The depicted curves have been filtered with a 4th-order lowpass Butterworth filter with 20 Hz cut-off. The filtered peak acceleration measured at the deck lies just below 8 g, while the measurements at the pan and cushion were approximately 2 g.

6. CONCLUSIONS

The sea trials showed substantial differences in performance between the various seats considered, demonstrating the importance of matching the seat to the boat.

By assessing the seat performance in response to each individual wave, the 'wave by wave' approach does not require testing in consistent sea conditions for sustained periods of time. Each test run produces a large number of individual SEAT values over a range of conditions, rather than one average value for a single, nominal condition.

This analysis method was not used, on its own, to determine the seat to be selected. Many factors must be taken into account in selecting a marine craft suspension seat including weight, cost, footprint, vertical space requirements, integration with controls, ease of adjustment and repair, robustness, and consideration of the severity of motions that are likely to be experienced regularly and in extreme circumstances. The ability of a seat to offer useful attenuation of shock and vibration in the environment in which it will be used is, however, a critical requirement.

7. REFERENCES

1. INTERNATIONAL ORGANIZATION FOR STANDARDIZATION, Earth-moving machinery — Laboratory evaluation of operator seat vibration, *ISO 7096,* 2000

2. INTERNATIONAL ORGANIZATION FOR STANDARDIZATION, Agricultural wheeled tractors — Operator's seat — Laboratory measurement of transmitted vibration, *ISO 5007,* 2003

3. RILEY, M. and COATS, T, Quantifying Mitigation Characteristics of Shock Isolation Seats in a Wave Impact Environment, *NSW/CCD-080-TR-2015/010 Naval Surface Warfare Center, Carderock Division*, 2015

4. NAVSEA TECHNICAL PUBLICATION T9006-AE-TRQ-010 Laboratory Test Requirements for Evaluating the Mechanical Shock Attenuation Performance of Marine Shock Isolation Seats, *US Naval Sea Systems Command*, 2019

5. WU, X. and GRIFFIN, M.J., Towards the standardisation of a testing method for the end-stop impacts of suspension seat, *Journal of Sound and Vibration , Volume 192(1), pp. 307-319*, 1996

6. GUNSTON, T. The sensitivity analysis of a suspension seat dynamic model, *PhD Thesis, University of Southampton*, 2002

7. LEWIS, C. and GRIFFIN, M.J., The implementation of an improved anthropodynamic dummy for testing the vibration isolation of vehicle seats. *Presented at the UK Conference on Human Response to Vibration, Buxton, Derbyshire*, 1998

8. INTERNATIONAL ORGANIZATION FOR STANDARDIZATION, Agricultural wheeled tractors and field machinery — Measurement of whole-body vibration of the operator, *ISO 5008,* 2002

9. FAIRLEY, T. & GRIFFIN, M.J., The apparent mass of the seated human body: vertical vibration, *Journal of Biomechanics, Volume 22, p. 81 – 94*, 1989

10. INTERNATIONAL ORGANIZATION FOR STANDARDIZATION, Mechanical vibration – Laboratory method for evaluating vehicle seat vibration – Part 1: Basic requirements, *ISO 10326-1,* 2016

11. INTERNATIONAL ORGANIZATION FOR STANDARDIZATION, Mechanical vibration and shock -- Evaluation of human exposure to whole-body vibration -- Part 1: General requirements., *ISO 2631-1,* 1997

12. HEALTH AND SAFETY EXECUTIVE, Whole body Vibration, *HSE Publication L141*, Health and Safety Executive, 2005

13. BRITISH STANDARDS. Minutes of the 21 February 2018 meeting of GME/021/06/03:

Mechanical vibration, shock and condition monitoring - Human exposure to mechanical vibration and shock - Whole body vibration and shock, 2018

14. HMSO. The Control of Noise at Work Regulations 2005

15. MARITIME AND COASTGUARD AGENCY. The Merchant Shipping and Fishing Vessels (Control of Vibration at Work) Regulations 2007

16. GUNSTON, T. and CLARKE, A. Development, test and selection of a shock mitigation seat for an all-weather lifeboat, *Proceedings of the 54th UK Conference on Human Responses to Vibration, Edinburgh, 24-26 September*, 2019

8. AUTHORS BIOGRAPHY

Dr Tom Gunston is the Principal Scientist for KINETIX, the CDG Coast Dynamics Group data acquisition and analysis systems. He is a Physicist with a PhD in whole-body vibration (WBV) and suspension seating performance and has specialised in managing WBV exposures for over 20 years, principally in fast boats but also in armoured vehicles, fixed and rotary-wing aircraft, industrial and agricultural vehicles and construction equipment. He was a Research Group Leader on human vibration exposure at the defence contractor QinetiQ and has been a Research Fellow at the University of Southampton and at the University of Sheffield. He is currently the Chairman of the British Standards committee on human exposure to vibration.
At the time of the work reported in this paper Dr. Gunston was an independent consultant contracted to the RNLI.

Andrew Clarke is the Compliance Engineer for the Royal National Lifeboat Institute (RNLI) within the engineering office. He has a bachelor's degree from Southampton University in Naval Architecture specialising in small high performance craft. Since graduation he has worked for the RNLI as a Naval Architect for two years and more recently as a Compliance Engineer. At the RNLI Andrew has been heavily involved in the whole-body vibration project and conducted numerous sea trials looking at vibration levels for crew.

CYBERSICKNESS AND EVALUATION OF A REMEDIATION SYSTEM: A PILOT STUDY

L. Andre and **R. Coutellier**, Naval Group Research, France

SUMMARY

Many questions arise regarding the use of virtual reality (VR) in the naval military field. This is particularly the case from a human factors' perspective, where system's usability is supposed to guarantee user's performance. Cybersickness, resulting from a sensory conflict between visual and vestibular systems, is one of the major limitations to the development of VR. Cybersickness can lead to nausea, oculomotor discomfort and disorientation. The major aim of the current study was to evaluate the efficiency of a remediation system for cybersickness. This system is designed to help remove the sensory conflict, thanks to miniature LED screens placed in the head mounted display (HMD). 18 subjects were confronted with a dynamic environment in VR, equipped with HMDs. Different physiological variables were measured during immersion. Every subject showed effects of cybersickness, starting, on average, after eight minutes of exposure, even if the system may reduce symptoms under certain conditions.

Due to confidential restrictions, the current article will not provide the systems industrial name, neither will it go into details about certain functionalities nor will it provide visual references to it.

1. INTRODUCTION

VR is a technology that is constantly growing. Defined as a technology that immerses a user in a virtual world through 3D LED displays such as HMDs [6], this technology was developed following the first presentation of the "ultimate display" concept, in 1985 [12]. The real outbreak was in 2014 with the start of sales of HMDs to the general public [14] that led to an infatuation for virtual environments. Since then, the number of scopes and applications of VR has kept progressing [14], from games or education to medicine, engineering and industry. Within Naval Group, the use of VR is oriented towards learning, training, simulations or ships and systems conception [2].
The different studies conducted under experimental conditions within Naval Group [2] have confirmed both positive and negative points about the use of VR. Firstly, this technology allows a variation of a large number of factors while limiting the cost. Secondly, it gives the possibility of observing users and their activity, which is important for human factors specialists. Thirdly, the possibility to put users in difficult situations is beneficial because this avoids taking unnecessary risks and putting subjects in danger.
However, there are still some negatives, such as the limits of the virtually recreated situations in relation to a reference situation (interactions between human beings for example), or cybersickness, to which will be devoted the next chapter.

2. CYBERSICKNESS

Cybersickness is a psychophysiological response to exposure to virtual environments [1]. The most widely accepted theory about cybersickness is the sensory conflict it creates [6], due to non-conformity between the dynamic interface that the eyes can see and the absence of movement perceived by the body. The first symptom review of motion sickness was done in 1964 [4], and many updates were made since then [5]. Symptoms can be divided into three categories: nausea, disorientation (postural instability) and oculomotor effects (eyestrain or blurred vision) (See table 1).

Table 1: Symptoms of cybersickness [5].

Nausea	Disorientation	Oculomotor
General discomfort	Dizziness (eyes open)	Headache
Fatigue	Dizziness (eyes closed)	Fullness of the head
Salivation increasing	Vertigo	Difficulty concentrating
Sweating	Disorientation	Eye strain
Nausea		Difficulty focusing
Stomach awareness		Blurred vision
Burping		

A number of factors can increase the risk of suffering from cybersickness [1]. Firstly, some of them are associated with user's characteristics, such as age, gender or whether or not the user has already performed the given task in the real-world. Secondly, some factors are associated with the VR system, like realism, system's calibration or its refresh rate. Thirdly, the factors associated with the task, like the adopted body position, the required head movements or the duration of use. It was reported that, for 60% of users, an immersion for 20 minutes in VR has a tendency to worsen cybersickness symptoms [9]. Finally, it is stated that passive motions are more likely to cause motion sickness than active motions [11].
Knowing that cybersickness affects a vast majority of users and that it reduces marine's training efficiency, it

was essential for human factors specialists in Naval Group to find ways of remediation to help them.
The simplest remediation method is the use of adaptation programs. Through various and regular exposures, users tend to become less sick as they adapt to the virtual environment [11]. However, the problem with adaptation programs appears when a user accustomed to VR returns to the real world, and need to readapt. There are other remediation systems, more intrusive, like a bone transduction headband sending electrical signals transmitted to the inner ear and allowing the brain to be deceived on variations of equilibrium. It is also possible to use a Wii Board to guide a user's movements. Virtually reproducing a user's movement in VR has also been studied, even if the results were inconclusive [7].
The system presented in the current study was developed by a French startup specialized in motion sickness remediation. The proposed solution was the recreation of an artificial horizon. This solution is supported by the literature as a mean to remove the sensory conflict [10, 13]. The startup's solution recreates a virtual horizon by placing 2 miniature screens of 9 LEDs in an HMD, which can be seen in peripheral vision. Thanks to an accelerometer, the LEDs would follow user's head movements and maintain a horizontal line parallel to the ground. The proposed solution should remove the sensory conflict, responsible for cybersickness, by synchronizing eye's and body's horizontal markers.
This was the first solution studied by Naval Group's human factors specialists in research and development, and it was chosen for being innovative and non-invasive. Indeed, it was important to find a way of remediation for cybersickness that does not disrupt the user during his immersion in VR.
The primary goal of this study was to evaluate the remediation system. The aim was to conduct a pilot study using an experimental approach. The study was conducted with a small sample in order to carry out a first impact analysis of this system.
The secondary goal was to find effective solutions to evaluate cybersickness between objective and subjective measures. In order to do so, it was necessary to create a situation of cybersickness under experimental conditions. Cybersickness is one of the major obstacles that hinder the development of VR and severely limits the performance of users suffering from it [1].

3. MATERIALS AND METHODS

3.1 PARTICIPANTS

Eighteen subjects (12 males and 6 females) were divided into three age groups (18-25yo, 26-50, 51+). They were randomly assigned to two conditions (with or without the system) with an equal number of males and females (4 males and 2 females) in each group. Participants were excluded if they were sick, pregnant, under medication or suffering from disturbances of balance or the inner ear. Any expert in VR (more than 5 hours of use), scared of roller coasters or scoring less than five at the Motion Sickness Susceptibility Questionnaire (MSSQ) [3] were excluded from the experiment. The MSSQ questionnaire was developed to evaluate the sensibility to motion sickness from childhood to adulthood. Moreover, participants were not informed that they were testing a system of remediation for cybersickness and did not know what the object of the study was.

3.2 PROCEDURE

Participants were asked to rate their condition on a scale between 1 "feeling very bad" to 10 "feeling very good" and to complete the Simulator Sickness Questionnaire (SSQ) [5]. This questionnaire derives from the Pensacola Motion Sickness Questionnaire (MSQ) [4], developed with the goal of evaluating motion sickness. The SSQ was readapted to measure only the symptoms linked to simulations, like cybersickness in VR. It is a succession of 16 states that the participant has to rate from 0 (no symptoms) to 4 (severe symptoms). After that, participants did the Romberg's posture test (the participant must stand with arms outstretched, close their eyes and maintain this position for 20 seconds. His oscillations are observed and reported with attention) before being equipped with the system CAPTIV© (sensors measuring electrocardiogram and galvanic skin response). Afterwards, they were fitted with the HTC Vive and informed of the fact that VR can generate nausea, and that therefore they were allowed to remove the helmet and stop the experiment at any time if needed. They were asked to verbalize as much as possible how they were feeling throughout the test. Then, the experiment began with a 5-minute session of VRSailing© (See figure 2), a simulation of a sailing ship. Afterwards, participants were asked to rate their condition again, and before starting the second game, IncellVR© (See figure 3), a roller coaster like environment. Participants were supposed to do 7 levels of this game, each level being of 2 minutes long. Participants were asked twice to rate their condition during the game and once more at the end, when the HTC Vive was removed. Then, they were given the SSQ again followed by the Romberg's posture test. In the end, an interview was conducted to record their impressions and feelings. Two hours after the end of the experiment, participants were asked to rate their condition and to express if any symptoms persisted.

3.3 DATA COLLECTION AND ANALYSIS

This study used a 2x2x3 between groups experimental design: two levels of gender (male and female), two levels of system equipment (remediation system equipped or No remediation system) and three levels of age (18-25, 26-50, 51+). Heart rate and galvanic skin response (GSR) were recorded via the CAPTIV L7000 system©.

Figure 2: VRSailing©.

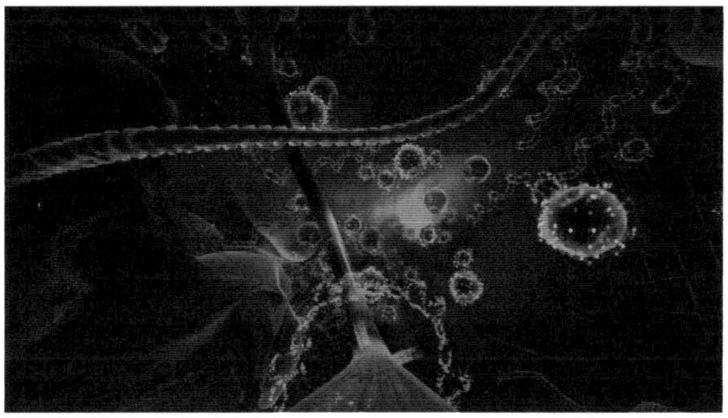

Figure 3: IncellVR©.

Statistical significance (showing the impact of one variable on another) was indicated by $p<0.05$, tendencies were also studied however, and corresponded to p between 0.05 and 0.1. Averages for condition rate with or without the system were studied, and the impact of the system on the different variables was measured thanks to parametric samples t-test. To evaluate the links between gender, age and the system and their impact on the variables, Anovas were conducted.

4. RESULTS

Fifteen participants out of eighteen reported symptoms of cybersickness, but all of them showed at least one effect of cybersickness (ECG, SSQ, Romberg's test score or GSR increase). Ten managed to finish all seven levels and eight did not, due to symptoms like nausea. For those who finished, four were equipped with the system and six were not. For those who stopped before the end, five were equipped and three were not. Among the six women who took part in the experiment, only one managed to finish the ride. Most of the participants found symptoms from level two of IncellVR©, approximately eight minutes after the beginning of the experiment, none reported feeling cybersickness during VRSailing©.

4.1 CONDITION RATE

Participants were asked to rate their condition once before starting, four times during the experiment and once two hours after. Results showed, for each of the groups, that the state of the participants deteriorates as the experience progresses and improves once the helmet is removed (See figure 4).

4.2 IMPACT OF THE REMEDIATION SYSTEM ON VARIABLES

The evolution of the variables in regards to the use of the system was analyzed using the parametric matched samples t-test. Tests revealed that the system limits the increase of the Romberg's test score ($t(16)=1.79$, $p=0.093$) as well as the disorientation dimension of the SSQ ($t(16)=1.98$, $p=0.065$), as it can be observed as tendencies in the Table 2. No significant variations were observed for other variables during the use of the remediation system.

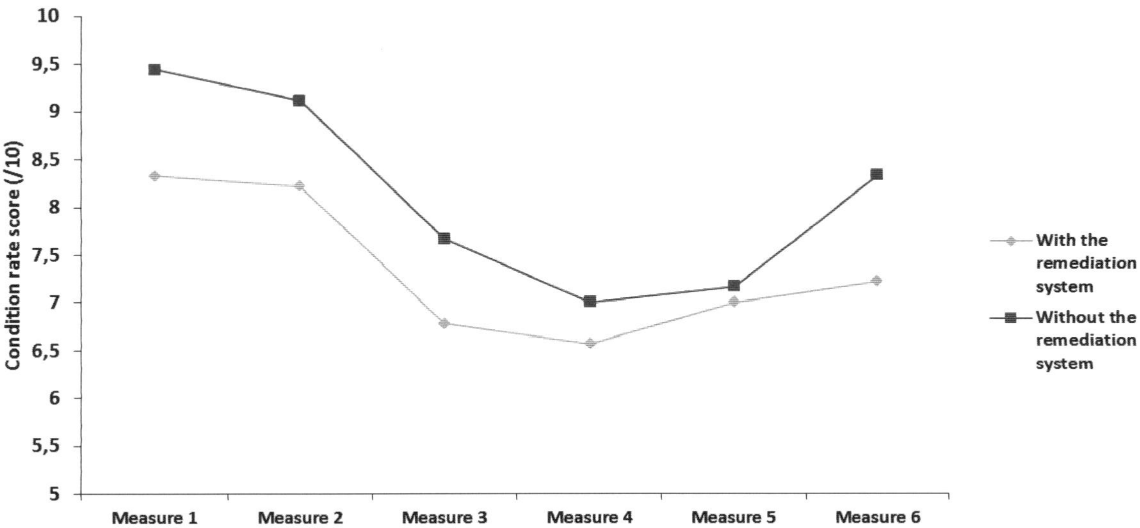

Figure 4: Evolution of condition rate depending on system equipment.

Table 2 : Matched samples T-Test.

Variable	*t*	*p*
Electrocardiogram	0.651	0.524
Romberg's test	**1.79**	**0.093***
SSQ	1.18	0.256
Condition rate during experiment	0.673	0.511
Condition rate after experiment	0.535	0.6
Nausea items of SSQ	-0.333	0.734
Disorientation items of SSQ	**1.98**	**0.065***

* Tendencies (0.05 < p < 0.1)

4.3 SIMULATOR SICKNESS QUESTIONNAIRE

Two-way Anova revealed significant effects of the interaction age*use of the system on the increase of the SSQ score ($F(2,16)=2.829$, $p=.098$). A Post-Hoc analysis showed that the remediation system limits any increase of the SSQ score for the younger group (M=4, SD=3 with the system, M=14, SD=3.606 without). No significant effects were noted about the other age categories (See figure 5).

4.4 MEN AND WOMEN DIFFERENCES

Some differences were noted between men and women's results. Firstly, only one woman out of six was able to finish the experiment, unlike men, as nine out of twelve of them managed to finish. Furthermore, Mann-Whitney tests were conducted to analyze links between variables and gender, and showed significant results for variables "levels done" ($U(16)=59$, $p=0.020$) and "success" ($U(16)=57$, $p=0.026$). Tendencies effects of the variable gender on the galvanic skin response were observed ($U(16)=55$, $p=0.083$). Indeed, women (M=2.239, SD=0.882) tend to sweat less from the hands, than men (M=4.592, SD=0.832). Since most women did not complete the experiment, the percentage of success and number of completed levels were lower. All other variables showed no significant results.

5. DISCUSSION

This study used two VR games involving vection (visual displacements) and all participants showed at least one of the usual cybersickness symptoms.

Regarding the remediation system, the aim was to evaluate its effectiveness in reducing the effects of cybersickness. Results showed that this system is effective in some situations, like to limit the increase in the SSQ score of young people. Further studies will be needed in order to better understand the system's performance in other conditions.

This was a pilot study, as explained in the introduction, and it was designed in order to get an initial idea of whether or not the system can reduce cybersickness. Even though the sample size of this study is relatively small, it provides an initial understanding of what could be expected if the population size is increased.

Concerning the condition rate score, it allows identifying a scheme towards the evolution of the state and making it possible to suppose that there is no impact in participants' perception of their condition from the use of the remediation system. However, this score is taken from the participants' own perception, and may therefore include biases due to the subjective evaluation. These results confirm the temporal evolution of the cybersickness symptoms, which are triggered after 8 minutes of use [15] (corresponding to measurement 3, see figure 4.).

Romberg's posture test and disorientation items of SSQ are related because they evaluate the same thing, namely postural instability. Despite the lack of correlation between these two variables, the results showed that the

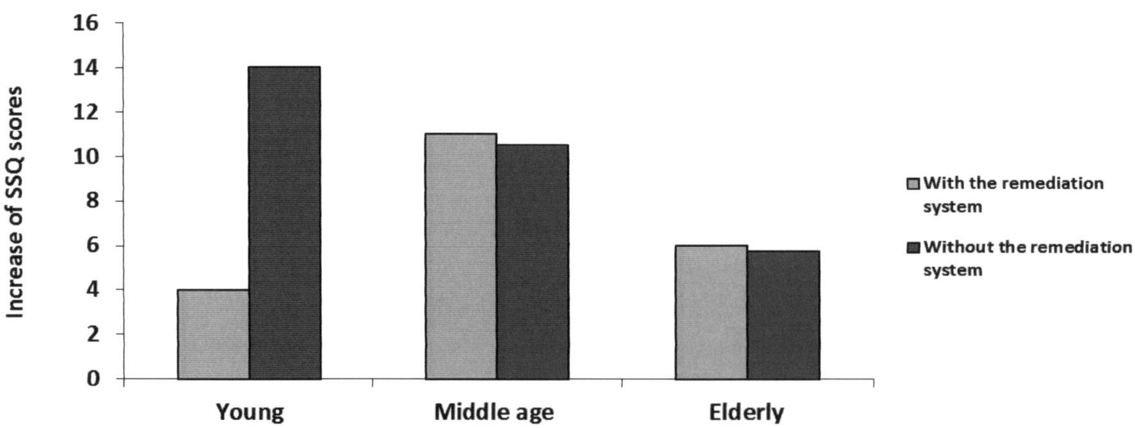

Figure 5: Increase of the SSQ scores depending on age and system equipment.

remediation system can restrict the upward trend of these two variables. The system would therefore have a positive impact on postural stability and the general

feeling of disorientation. The potential existence of an evaluation bias by the participants may be explained by the fact that the two variables are not of the same nature (one being objective and the other subjective).

The SSQ questionnaire has been used in many studies and its validity to measure cybersickness has been proven. The present study's data on the SSQ scores showed that the score increases for the entire population during the experiment. It is important to note that, in agreement with the literature [1] and when results are isolated without the system, young people report feeling more symptoms than the middle ones, themselves experiencing more symptoms than the elderly. When comparing results with and without the system, the conclusion that can be made is that the system seems to only affect the most sensitive population, the young [1], and has no effect on middle-aged or elderly participants, who are less susceptible to motion sickness. These results suggest that the remediation system studied in this paper reduces the symptoms of the population who is very susceptible to show effects of cybersickness (younger people, with antecedents of motion sickness).

Several studies [1, 6] presented women as more sensitive to cybersickness, this is confirmed in the present study. Concerning the lower electro dermal response score for women, a physiological explanation can be provided. A study [8] was conducted about sweating patterns in men and women and concluded that people of small body size (usually women) are more dependent on heat loss through increasing circulation than by sweating, unlike people of high body mass (usually men). However, it is important to note that, for women, there was only one participant in each condition which can be considered as a non-representative sample.

6. CONCLUSIONS

The results presented in this paper showed that the system is efficient to reduce symptoms of cybersickness for the most sensible population. However, those results have to be taken with a grain of salt given the limitations of the study. As a pilot study, it is encouraging and calls for more elaborated studies in order to understand how the remediation system could affect the appearance of cybersickness in a training context for example. But this study also shows the need for better understanding on the relationship between the notion of cybersickness and the existing method of measurements, since the remediation system seems to affect differently some of the measurement (i.e., objectives vs subjectives). As the use of VR technology is still growing and its usefulness can be asserted for different field, remediation systems will help eliminate the main obstacle of its use, the cybersickness.

7. REFERENCES

1. Barrett, J. (2004). Side effects of virtual environments: A review of the literature (No. DSTO-TR-1419). Defence Science and Technology Organisation Canberra (Australia).

2. Coutellier, R., Maïs, C. (2019). Evaluation Facteurs Humains des systèmes innovants en Réalité Virtuelle pour le domaine naval militaire. Communication présentée à EPIQUE, Lyon, 9-12 juillet 2019

3. Golding, J. F. (1998). Motion sickness susceptibility questionnaire revised and its relationship to other forms of sickness. Brain research bulletin, 47(5), 507-516.

4. Kellogg, R. S., Kennedy, R. S., & Graybiel, A. (1964). Motion sickness symptomatology of labyrinthine defective and normal subjects during zero gravity maneuvers. AEROSPACE MEDICAL RESEARCH LABS WRIGHT-PATTERSON AFB OHIO.

5. Kennedy, R. S., Lane, N. E., Berbaum, K. S., & Lilienthal.,M. G. (1993). Simulator sickness questionnaire: An enhanced method for quantifying simulator sickness. The international journal of aviation psychology, 3(3), 203-220.

6. LaViola Jr, J. J. (2000). A discussion of cybersickness in virtual environments. ACM SIGCHI Bulletin, 32(1), 47-56.

7. Loup, G., & Loup-Escande, E. (2019). Effects of travel modes on performances and user comfort: a comparison between ArmSwinger and teleporting. International Journal of Human–Computer Interaction, 35(14), 1270-1278.

8. Notley, S. R., Park, J., Tagami, K., Ohnishi, N., & Taylor, N. A. (2017). Variations in body morphology explain sex differences in thermoeffector function during compensable heat stress. Experimental physiology, 102(5), 545-562.

9. Regan, E.C., Price, K.R., (1993a). Some side effects of immersion Virtual Reality in a population of firefighters. APRE Working paper 9/93.).

10. Rolnick, A., & Bles, W. (1989). Performance and well-being under tilting conditions: the effects of visual reference and artificial horizon. Aviation, space, and environmental medicine.

11. Stanney, K. M., & Kennedy, R. S. (1997). The psychometrics of cybersickness. Communications of the ACM, 40(8), 66-68.

12. Sutherland, I. E. (1965). The ultimate display. Multimedia: From Wagner to virtual reality, 506-508.

13. Tal, D., Gonen, A., Wiener, G., Bar, R., Gil, A., Nachum, Z., & Shupak, A. (2012). Artificial horizon effects on motion;sickness and performance. Otology & Neurotology, 33(5), 878-885.

14. Tokey, K. T. (2018). Conception et évaluation de l'expérience utilisateur en environnement virtuel immersif (Doctoral dissertation, Ecole nationale supérieure d'arts et métiers-ENSAM).

15. Wang, G., & Suh, A. (2019). User adaptation to cybersickness in virtual reality: a qualitative study.

9. AUTHORS BIOGRAPHY

Lucile ANDRE holds the current position of human factors specialist at Naval Group Research, in France, in the human factor's department. She is responsible for evaluating new technologies in a human factor's perspective and her main area of focus is human-systems interactions. She graduated from Aix-Marseille University in 2019 and holds a degree in psychology with a major in human factors.

Rozenn COUTELLIER is a Human Factors specialist at Naval Group Research. She is in charge of the installation and operationalization of the Sea User Lab. Her previous experience includes both HMI ergonomics and organizational aspects for user impacts: analysis, design and evaluation in a User Centered Design approach.

ANOTHER SERIES OF UNFORTUNATE EVENTS: FAILURES IN SITUATIONAL AWARENESS

M.J. Cook, BAE Systems, UK

SUMMARY

In the Human Factors literature there is now a very sophisticated understanding of human factors engineering and how control teams function. In particular the differentiation of technical and non-technical skills and their role in supporting effective control team performance has been outlined. Many of the significant changes have come from the aviation industry and other application domains, like maritime, are attempting to catch up. However, there is a dearth of detailed information in maritime because unlike aviation the use of cockpit voice recorders and other detailed recording of platform behaviour is not necessarily mandated or recorded. Training simulators offer a good opportunity to explore these issues but in the absence of an adequate quality of training evidence, accident reports provide an insight into the contribution of human factors issues, technical and non-technical, to incident development and accident sequences. As pointed out in earlier papers commercial shipping is not immune to these issues associated with human performance and technology interactions, with incidents such as the Viking Sky and Savannah express indicating analogous issues. To categorise events such as those associated with the USS Fitzgerald, USS McCain and HNoMS Helge Insgtad as a failure of human operations is a statement of the obvious and not a guide to future improvements required to prevent recurrence.

NOMENCLATURE & TERMINOLOGY

Human Factors – For this paper it describes the full range of variables that might be considered in a socio-technical system from cultural to performance, including cognitive and physiological which shape performance outcomes and are related in some way to the functional performance of the human component of the system.

Naturalistic Decision Making – a term used to describe how human decision making processes occur.

Situation Awareness – A term used to describe an individual or a team's awareness of their current operational environment, their equipment status and the resource availability to solve problems.

Sociotechnical system is a combination of the human and technological components that allow system capability to be achieved. This implicitly acknowledges same-same interactions between humans or machines, and in addition human-machine interactions, where the influence is direct, or indirect.

1. INTRODUCTION

Through an examination of accident reports and the consideration of human factors issues arising it is argued that Human Factors Integration Programmes may not be applying sufficient depth of analysis on the critical issues of multi-operator command and control systems on-board surface ships. It is suggested that this problem may have arisen because systems engineering programmes have dismissively pushed the detailed human factors analysis towards the end of the platform delivery and programme completion without due regard to the problems that this creates. By comparison in aerospace programmes the importance of advanced control systems (albeit now largely used by a single operator) has made use of Model Based Systems Engineering models and rapid prototyping to simulate control environments, so the a more advanced examination of system-team interactions can be examined in greater breadth, usually represented as scenarios. In more complex multi-crew platforms the role of the total system performance (involving systems and people) especially with regard to multi-crew synergy is largely underwhelming in maritime environments. The persistence of accidents with a non-technical skills basis suggests that this must be addressed and as early in the design lifecycle as possible.

2. SITUATION AWARENESS

Situation awareness is a term originally coined by Mica Endsley in the USA. Situation awareness is both a structural and a functional description of cognition, human information processing. Situation awareness is time bounded because it is dynamic and it can change over time. There are three states of situation awareness. The first state is basic and reactive, it is described as perceptual situation awareness and it reflects a mental state where processing may be more effortful and less linked to mental models of the system behaviour. The second state of situation awareness is comprehension and it reflects a genuine understanding of the system, the process and where the total system will be in the future. The system in this sense is everything outside of the operator and can include the world (environment), the plant under control and the models of system behaviour expected. The final situation awareness state is projection, where the model of possible futures is activated and the operators may be second guessing what they have to do. Working in the projection state where everything is known and understood, and a selection of responses are available is preferable. However, if an unexpected or unplanned for event occurs the operator(s) may be forced back into making sense of the on-going events or worse still, simply reacting to the events as they happen.

Consider the Savannah Express approaching the inner harbour when the engines failed and the ship was no longer under control. This unexpected event required that operators react to on-going events and they were unable to regain control before the collision with the linkspan bridge. What had happened is that the crew and the pilot had failed to foresee and understand the possibility of engine failure in confined waters, such that they failed to predict or project the future consequences. Sometimes events and their relationship to outcomes are spread out in time but the pre-emptive or corrective action needs to occur much earlier, as the late awareness of the event makes the attempted corrective action at that time ineffective. Knowing one is going to collide at a late stage is not helpful because the attempt to steer or slow the boats progress cannot be immediate in effect.

3. ACCIDENT EVIDENCE

Grech, Horberry and Smith [9] identified issues with a lack of situational awareness as a significant contributor to many maritime accidents, evidenced by analysis of accident investigation reports (supported later by Hetherington, Flin and Mearns [12]). Fundamental to this view is the acceptance that maritime control rooms, on the bridge of ships, in engine control rooms, or even ashore in remote control is based on a socio-technical system. The concept of Situation Awareness is a very useful and a valuable vehicle for understanding performance failure in these kinds of dynamic situations with its three level analysis [4, 5]. As noted already operators responding to events as they happen are often lagging behind event development and may struggle to comprehend the significance of the events they currently perceive. A typical error which may occur when situation awareness is limited is incorrect selection of the applicable Emergency Operating Procedure (EOP) because of failed comprehension, which results in delays to the correct actions or worse still actions that compound the events occurring. It is important to remember that one can fail to execute the correct actions or one can fail because the time course of events outstrips the execution of actions. Time is a critical factor as it modulates the risk.

Human beings have evolved to process information relatively quickly and identify patterns of events that indicate risks. Humans are not necessarily good at identifying in real-time sequences of events where an action is required earlier to pre-empt a later problem. A fact that is obvious in maritime accident sequences, where an earlier decision is disqualified by later events. For example, a change of course to enter harbour earlier may occur which puts the vessel at risk of grounding and this is more likely where the risk of grounding is qualified by tide times, such that the risk is only obvious at certain tides visually. A major risk, both in aviation and in maritime operations is centred on transitions or changes to operation. Where ships enter a strait it may be more or less obvious that there are various risks. In the straits of Hormuz the risk of terrorist attack, as occurred for the USS Cole (12th October 2000) when attacked by a small fiberglass boat containing C4 explosive, is an example of how focus or belief can make ships vulnerable. Risks can also occur with increasing traffic density, even where the intent of other maritime operators is not malign.

The vast majority of accidents occur as a result of failures in perception situation awareness and this is evidenced by the failure to start taking appropriate actions to pre-empt the future hazard realisation. Once the significance of events has been comprehended the successful prediction of outcomes and required responses become obvious, a fact that is implied by the proportional accountability that identifies 75% of accidents with initial perception failures. Rothblum [16] suggests that between 70-90% of collisions at sea involve some form of human error as a contributor, the exact figure is inconsequential but the magnitude is. It is important to note that human agency is critical to the operation of maritime vessels and so the contribution to accident occurrence should not be surprising. The problem is that current proposed methods of investigating human errors may not reveal design-based failures on human errors as some commentators suggest [1].

4. USS McCain & USS Fitzgerald

Applying the concept of situation awareness to the collisions of the USS John McCain and MV Alnic MC, the USS Fitzgerald and MV ACX Crystal [3] and the KNM Helge Ingstad with the oil tanker SOLA TS [2], supports the view that situation awareness had failed. See also [14].

If the McCain and Fitzgerald's crew had effective situation awareness they would have contacted local traffic on VHF to warn them of their presence and ensure the appropriate distance and course was kept. It is possible that the final failure, and the collisions, could be traced back to earlier decisions such as changing course management at a busy and inappropriate time, not ensuring that senior officers were on deck given the risk. Or, the failure in situation awareness may have been a result of the bridge equipment design that lead the crews to believe that they sufficiently aware with Automatic Identification Systems (AISs). There is a lot of newer technology such as AIS, and ECDIS/WECDIS charting systems that might be assumed to support effective de-confliction of traffic.

The analysis presented here identifies team or bridge situation awareness as an emergent property of human activities e.g. sensemaking [19], created from information that they receive. Thus, in protecting against such failures in future it requires more than an additive analysis of behaviour because of the non-linear nature of the interactions between people, technology, operational

guidance and training. In the simplest terms the information available is not the information being used to inform decision making and problem solving.
It can be argued that the failures and situation awareness describe the failed responses to equipment failure like that seen in the collision of the Savannah Express and the portside bridge construction in Southampton, UK. This suggests a failure by designers to appreciate that information availability does not guarantee transformation into knowledge, and the operator's tendency to resort to using poor seamanship in cases where such incidents occur is not helpful. This does not exclude the possibility of loss of situation awareness through relatively simple equipment failure as has happened with electronic charting systems or GPS inputs to mapping/navigation systems.

5. COLLISION ANALYSIS

The reports on the collisions of the USS McCain and USS Fitzgerald, in separate incidents, with commercial traffic make for sobering reading, and the speculation and conjecture that followed the publication of accident reports raises additional concerns. And yet collisions per se are not solely found in U.S. operations, the Royal Norwegian Navy lost the Helge Ingstad in attempting to avoid a collision. The similarities of some of the event context should be a cause for concern because of the frequent operation of military ships in littoral waters supporting land campaigns, the narrowness of straits and the limited depth available which can make operations challenging, as avoidance is not always a sound option. Added to this the presence of other independently operating traffic creates a challenging operational environment. However, many would argue that such challenges are expected and crews should be familiar with the real-time safety critical decision-making expected of them. It is argued that surface ships and other maritime craft are likely to face the same challenges that impacted aviation in the last 40 years. The electronic and computational systems the crews now use have radically altered the relationships between the crew members and as a consequence performance has in some cases deteriorated. In some cases crew numbers have been reduced but this is only one of a number of potential changes with the introduction of advanced technology.
The challenge has not been in absolute crew numbers but in the crew's understanding of their role, which is in part defined by the technology. In some cases, designs have largely focussed on individual roles and workstations, failing to look closely enough at the integrated team performance. The completion of the initial general arrangement may simply be a shopping list of people and things, without a due consideration or relationships between the crew's individual roles or the synergies in system operation. These early decisions cast a long shadow over the design development because they create inertia or momentum that makes the perception of future design decision as both risky and costly to reverse.

This casts doubt on the effectiveness of the methods applied to multi-crew system designs in the maritime environment where the commercial organisations are gradually becoming conscious of the human element in maritime operations. Previously it has been suggested that with increasing automation and use of advanced technology the scrutiny applied to the human role needs to increase, not decrease, as the margins for failure in the event of reversionary modes of operation are more finely balanced. Automated systems and advanced technology has an impact in shaping behaviour through its presence and in the way the expectations of the crew are driven in routine operations. In reversionary states the same performance challenge remains but it falls upon the operator to execute with an unfamiliar reversionary mode of operation.

6. DISCOVERY

The piecemeal transformation of the command and control environment has been a concern in military circles which has been enhanced by awareness of the network enabled revolution. Network centric warfare (Network Enabled Capability) has promoted an awareness of effective communication in military systems, the human-system interface, human error and more recently the people-systems-technology integration issues in Human Factors Integration plans. The problems with electronic charting systems, their safe operation and use is one of many surprising issues for sea farers and designers. The design challenge is that analysis of the problem is a multi-variate multi-layered problem in which the human element, the operational documentation, the human factors engineering and the understanding of system operation via methods such as task analysis are rarely completed for an individual operator. The effort and the impact of systems integration late in the design lifecycle are weak at best.

As noted already the last forty years have seen very significant changes to the civil flight deck in aviation, with changes that broadly mirror those now starting to occur in maritime systems. The crew size has been reduced, increasing levels of automation have been introduced, the crew's role is now shifted to be largely as a monitor of systems, and the complexity of systems makes the diagnostic analysis of system behaviour problematic, and poorly supported. There have been stern critics of the mainstream human factors analyses because of the tendency to produced negative attitudes towards human operators, when in fact the human operators may be seen as victims of the slow residualisation of the operator's roles and a failure to address human-centred design issues. The ironies of automation were identified long ago but there is still an absence of a solution to the problems.
This philosophical position with regards to the operator may be a secondary consequence of safety cases focussed on design risk, which is perceived as more easily recognised and quantifiable at lower levels in

system terms. The human element is often perceived as difficult to define or costly to pursue, so strategies are adopted to eliminate the human, even though the system cannot embrace all the necessary capability required. Part of the problem with the elimination of the human operator is the issues of unexpected or unplanned for events which are outside of the scope of the automated responses because then all that is left for the operator is the responsibility for the accident, after a defect or unexpected event. The focus on sunny day scenarios for many systems is an inherent weakness in their evaluation. However, such sunny day scenarios afford a much healthier prediction of performance for the automated systems with a much reduced problem set, however, what remains for the human operator is the unexpected and unplanned for events.

It is important to note that unplanned for events can occur in a number of guises. The failure of the sensors to detect oil in the tanks on the Viking Sky because of a large roll component meant that operators unexpectedly had to deal with engine shutdown. The same system protection approach lay behind the Savannah Express's engine shutdown, when sensors failed and tripped a protection. Seafarers will recognise this as a trait indicative of modern control systems which have the minimum barriers to defence which fail to accommodate all operations. The Boeing 737 Max failure is equally an example of sensor failures causing operational difficulties, so aviation is by no means immune to such issues.

Those advocating automation have by and large used the dictums that come from the Functional Allocation literature that is derived from the early work of Fitts [6]. It is perhaps interesting to note that some like Hancock and Scallen [11] have suggested that a dynamic and adaptive approach to function allocation be considered, without considering the additional complexity that this pre-supposes in the relationship between operators and systems. Or, considering the interface by which this complex relationship is to be managed. Others, such as Sheridan [18], seem to have abdicated responsibility for function allocation suggesting that we consider letting the evolving machines show us what they can achieve. The obvious difficulty is that in modern safety-critical systems one needs to demonstrably make a case that the system design is safe, as far as is reasonably practicable and commensurate with the consequences of accidents. It is disappointing that some seem to disregard the design arguments associated with synthesizing human and machine performance, replacing it with the irrationality of what will be, will be.

Sheridan [18] is notably critical of human-centred design arguing that it is fraught with inconsistencies in definition and application. It appears that the difficulty with quantifying the human-machine relationship is used as a justification for leaving it out of considerations. In addition, the models applied to human decision making, either Naturalistic Decision Making or Ecological (intuitive) design, are rejected by Sheridan as incompatible with normative decision theory, suggesting that perhaps designers do not need to consider the human perspective on information processing. The failure to acknowledge the importance of the human mental model of systems under control may be in part a reason why so many automation surprises occur and accidents sequences progress so far.

The counterpoint to Sheridan's view is an earlier paper by Fuld [7] who suggests that formal function allocation process is itself a design that has failed to consider the user's tasks. Unfortunately, Fuld stressed more local integration of human factors in detailed design activities and less emphasis on increasing the global authority of human factors over the design team. The tendency to apply human factors in an analytical way to individual roles and to fail to use human factors in the wider systems integration, as a socio-technical system was typical at that time. The attitudes to the crew in modern applications of human factors demonstrate greater insight into the way that technology shapes operators behaviour and interactions between operators in multi-crew systems, in that context function allocation should become a more experimental enterprise. However, the dominant force in complex systems is reducing the manning budget and the dependency on human operators. As such operators are characterised as error prone, low reliability elements of the system, which are better replaced by automated systems. At the same time the fragility of automated systems and their propensity for failure are not considered. As Leveson () has pointed out automation can produce highly sophisticated responses that give rise to the wrong outcome because automation is not as aware of the context as the human operator might be.

The difficulty of integrating human and machine intelligence was well demonstrated by an early paper which discussed the ways in which automation might become a better team player (Klein, Woods, Bradshaw, Hoffman & Feltovich (2004). The authors compared the human-machine relationship to that of human-human partnerships, and what is interesting is the way in which advanced technology can disrupt the human-human relationship in multi-crew environments. After the introduction of advanced technologies in the civil cockpit the rapid growth of crew-resource management and later the advent of non-technical skills training, can be strongly identified with issues that arose in the 1st and 2nd generation of automated cockpits. At the same time, figures like Eduardo Salas and associates were publishing extensively on the subtlety of team performance and the way in which it supports higher levels of team performance, with better outcomes and reduced completion of error sequences leading to accidents.

Klein et al.'s (2004) paper identified challenges in human-machine synergy that related to notifying partners of failure to complete actions, being mutually predictable (modelling intentions and actions), agents must be directable, agents must manage the dialogue with human operators, agents must interpret behaviours and intentions, agents must be able to engage in negotiation, support planning and autonomy should be collaborative, agents must be able to participate in managing attention and all the resource costs should be managed across the human and machine elements of the team. The majority of these activities require a sophisticated dialogue between the human and machine intelligence embodied in the system.

It is clear that the interface can embody elements of these processes with current technologies. Attention getting can be based on alarms and warning systems, although understanding the priority of alarms and warnings may be contextualised in a way that an automated process may find problematic. In terms of building a safety case the alarms and warnings philosophy applied to system design may involve prioritisation of tasks that must be completed immediately, within a short space of time and which require discriminative responding. The difficulty is that some tasks could be completed automatically by automated systems, but the differential responding in some instances (exceptions) may be difficult to encode and costly to verify for software. Automated systems currently are not self-contained, requiring sensors and communication pathways to determine the system state or environmental situation, and these external links can fail. The logic and delivery of robust supporting systems to automated analyses is challenging, and when it fails, making sense of the failure may not be immune to additional confusion.

7. SOCIO-TECHNICAL SYSTEMS

In the 1990s the term socio-technical systems was introduced to describe the way in which social issues influenced performance with a seminal book by Jirokta & Goguen [13]. Interestingly, socio-technical systems awareness arose when many of the large scale IT systems failed to deliver the performance or capability that was expected of them. In an early paper, Grudin [10] recognised that development may fail because of a simple-minded approach to systems development which equated multi-user application programs and multi-user computer systems, and the analogy between multi-user applications and single-user applications. This can be applied to the thinking about command systems where attention is focussed on the individual operator's roles and their workstation design. There is nothing synergistic about design that addresses the team functioning.

Grudin's [10] insights about the short-falls in understanding appear to be replicated today in system design for multi-crew systems, such as ship's bridges, where focus on individual roles and workstations is equated with overall effective system design. This assumes that there are no benefits in terms of synergy from multi-crew cooperation, or that accident sequences cannot arise because of misunderstandings between crew members in the joint operation of systems. The use of accident sequences that focus on root-cause analysis miss the point that accidents can be socially constructed events, arising from failed interaction. This is the essential paradox which exists between design and maximising safe operation. These two activities should be using the same evidential basis but they do not.

This early work was at a time when socio-technical systems were also described as CSCW (Computer Supported Cooperative Working) which was in the ascendancy after Irene Grief (1988), together with Paul Cashman, had created the term as way of describing concerns associated with supporting multiple individuals working together with computer systems. Such systems exist in Nuclear Power Stations, Chemical Process Plants, Control Rooms on Ships, on the flight deck in Commercial Aviation, Air Traffic Control and many other control environments, where computational support is afforded for the management of systems and processes. Ironically, Fitt's list which was generated in 1951 for applications in Air Traffic Control (ATC) or now also known as ATM, where the M is management, is a system which needs and benefits from a socio-technical systems approach.

Although the focus in this analysis is real-time time and safety Critical Applications, CSCW has been applied to information systems in healthcare, incident management, and many other areas where information is collected and interpreted to support decision making. The evidence from this research strongly supported the idea that socio-technical considerations were important in accident sequences and could not be captured by traditional methods focussed on human error and individual component failure [15]. The idea that factors such as cultural differences could enable accident sequences was considered possible and worthy of further investigation [8]. A good example of where this adds value to any analysis is presented by Salmon, Walker and Stanton [17] who analysed the Air France 447 flight which crashed in 2009. As Salmon, Walker and Stanton [17] note an equipment failure, in terms of the pitot tubes feeding the altitude and airspeed systems, failed after freezing over but the subsequent response of the crew to the failure was problematic. It was argued that the socio-technical system failed, not the individuals, in that situation awareness was lost across the system, resulting in the subsequent stall and crash. The same concern could be applied to many advanced control environments as noted above, defects do occur and systems do not always respond as expected station in control is not transferred completely, sensors are discordant between workstations, and so on. Part of the human operator's role is to diagnostically identify "odd" events which suggest problems with the system.

8. HUMAN FACTORS EVENTS

Wiegmann and Shappell [have provided a model of applying human factors to accidents, the HFACS (Human Factors Analysis and Classification) system, which can arguably be applied to the collisions involving the USS McCain and USS Fitzgerald. Organisational influences, unsafe supervision, preconditions for unsafe acts and unsafe acts themselves can all be potentially identified in the reports on the collisions recorded in the available literature in the public domain.

It was clear that the training on-board the USS Fitzgerald was not adequate and this was reflected in number of areas such as knowledge of the rules of the sea, assessment of the situation as potentially hazardous, and knowledge of the use of equipment on-board which supports safe navigation e.g. the radar and the automatic identification system (AIS). In terms of task execution it was argued that the ship was not moving at a safe speed, that manoeuvres were not conducted when required, and that other ships were not notified to allow action to be taken in a timely manner. The failures in basic knowledge, use of equipment and in situation assessment could be individual failures but these were clearly compounded by failures by the ship's crew to self-correct through challenge. A key point in Navy Doctrine is the principle of supervision and correction, such that any team member that identifies issues can raise them and challenge. Similarly, supervisors identifying problems should correct the behaviours of those sub-ordinates that they see executing inappropriate behaviours. The British Navy operates a SCAR (Select-Check-Act-Review) principle which helps to ensure that any executed action is reviewed in terms of consequences and correctness. These principles should work to reduce erroneous behaviour if followed. However, social loafing and diffusion of responsibility can undermine these corrective actions if a more casual approach is applied.

It was noted that the bridge team and Combat Information Centre (CIC) did not communicate and coordinate effectively to secure a safe passage, to respond promptly to situational challenges, and apply leadership to rectify issues was absent. Thus, it seems that errors and violations, which are identified as unsafe acts in HFACS, occurred. However, this was prefaced by substandard conditions of operators and substandard practices of the same operators. The individual failures were compounded by unsafe supervision, inadequate supervision, inappropriate operations, failures to correct problems and supervisory violations (absence of leadership). Although it is attractive to perceive this through the individual responsibility of operators the evidence suggests a more collective failure of the kind that would be termed non-technical skills or organisational failure, according the James Reason's model.

Non-technical skills are those skills which are not bound to technical competencies but which enable effective functioning of multi-crew systems. Leadership, teamwork, decision-making, situation awareness, communication and problem solving are all examples of non-technical skills competencies. There should have been effective leadership to prioritise responses to challenges, such as the operational environment, and to guide effective responses when elements of the process failed. Teamwork, in the form of shared and collective responsibility should have motivated individuals to complete their assigned roles and to challenge others when they created opportunities for failure. Decision-making and situation awareness should have helped to ensure a correct preparatory response to narrow lanes of traffic and traffic density, enhanced by leadership altering in response to surrounding traffic to avoid potential collisions. Communication was clearly lacking between the bridge and CIC, but it was clearly a general problem in that crew members failed in their own roles and failed to challenge those around them. Finally, when the collisions were about to occur the crews made no attempt to alert or warn those around them, to minimise the impact.

In the case of the USS McCain there were similar issues. The training was clearly not adequate in the use of the on-board systems and errors of operation were compounded with the situation which led to the collision. Almost all the relevant crew on-board and on the bridge, including the most senior officer, lacked the relevant knowledge to operate the systems. Again the crew appeared to have seriously under-estimated the degree of readiness and the need to alert others in terms of warning sounds, waiting until it was too late. It is interesting to note that conflicting views between the senior officer and the navigations, operations, and executive officer were not effectively resolved and safer actions were ignored. There was a leadership failure in terms of providing effective supervision. Unplanned changes in control were instituted without consideration of human error or establishing control when changes were made. There was a misunderstanding between those present about what was actually taking place. There was not the critical cross-checking of the situation when the loss of steering was reported. Again, ineffective leadership, poor team work, ineffective communication, poor situation awareness, faulty decision-making and poor problem solving were all present.

Behind these issues is a potential false-belief that automation and enhanced control systems reduce the need for multi-crew operations to engage the participants in thinking ahead. That is what defines situation awareness, the ability to think faster because information has been perceived, comprehended and can easily be used to predict the future, through projection and prediction of future events. Entering into difficult circumstances, busy traffic environments and narrow straits, were grossly under-estimated, the context was set aside. The subtlety of crew to crew interactions is significant. Many control environments have a very structured dialogue between those present and it reflects the cognitive breakdown in responsibilities, and this can be disrupted by new technology. For example, the introduction of Automatic Identification Systems (AISs) made crews less likely to engage across VHF with other ships, in the belief that they could see others and perhaps others could detect them. Senior crewmen ask more junior crew to review the instruments and information sources, to report back. The senior crewmen are one step removed, forming situation awareness (the bigger picture) and responding to the reports of junior crewmen. Once the initial picture is built it is critiqued to establish a sound course of action. All crew members are

participants in this process and can shape the final conclusions reached. This distribution of workload and actions has intrinsic value in ensuring the best possible processing of available information. The presence of shared screens and displays makes some believe this is no longer necessary.
Did junior crewmen assume that the current radar image was available to all the critical parties? Did they assume the context was not so challenging because senior crew knew of the traffic and geographical information. Was there bystander apathy? Or, could it have been diffusion of responsibility across the command teams? In aviation there are many accidents in which lives were lost before the impact of technology on cockpit situation awareness was fully appreciated. It is easy to lose the subtlety of dialogue skills that support enhanced situation awareness and subsequent decision-making. And yet, the synthetic aspect of design may not be well incorporated in the organisation operating platforms or establishing standards for training. The roles of individuals and the failure rates of components are a long way from the factors that contribute to modern performance failure. The situations that give rise to such failures are currently given limited analysis and examination in designs which focus on individual roles, workstations and do not assess the socio-technical impact of design thoroughly.
This analysis is not definitive; it is a foil, a straw man. The authors want to challenge designers to demonstrate they understand the problem, have methods to address the problem, the methods can be applied early enough and the desired outcomes can be achieved. Future work will consider just how early the design work can start and what form that design work might take, as well as what resources are needed to support the systems integration issues of command and control environments.

7. CONCLUSIONS

Recently a body of work has established that non-technical skills play a vital part in ensuring successful team execution of command and control actions. Recent accident sequences appear to corroborate this but the technology in the maritime environment has changed. It is concluded that non-technical skills may be influenced and undermined by technology change. This may be exacerbated by the tendency to conduct systems integration late in the design lifecycle when empirical validation is unlikely to result in design change, if indeed empirical analysis is funded.

8. ACKNOWLEDGEMENTS

BAE Systems has very generously helped to support this work but the views expressed are wholly those of the author and based upon the referenced sources identified. This paper was submitted for clearance prior to release and contains no reference to sensitive material and makes no use of any material not already available in the public domain.

9. REFERENCES

1. Celik, M., & Cebi, S. Analytical HFACS for investigating human errors in shipping accidents. *Accident Analysis & Prevention*, 2009, *41*(1), 66-75.

2. Defence Accident Investigation Board Norway. Preliminary marine accident report – Collision between the frigate KNM Helge IOngstad and the oil tanker Sola TS on 8th November 2018, outside the Sture terminal Hjeltefjorden in Hordaland county. 2018 Accident Investigation Board Norway.

3. Department of the Navy. Enclosure (1) Report on the collision between USS Fitzgerald (DDG 62) and motor vessel ACX Crystal. Enclosure (2) Report on the collision between USS John S. McCain and motor vessel Alnic MC. Department of Navy, 2018.

4. Endsley, M.R. Toward a theory of situation awareness in dynamic systems. Human Factors, 1995, 37(1), 32-64.

5. Endsley, M. R., & Garland, D. J. (Eds.). *Situation awareness analysis and measurement*. 2000, CRC Press.

6. Fitts, P.M. (Ed.) Human Engineering for an effective air navigation and traffic control system. 1951, Washington D.C.: National Research Council.

7. Fuld, R. B. The fiction of function allocation. *Ergonomics in Design*, 19993, *1*(1), 20-24.

8. Gorman, J.C, Cooke, N.J., Salas, E. and Strauch, B. Can cultural differences lead to accidents? Team cultural differences and sociotechnical system operations. *Human factors*, 2010, *52*(2), 246-263.

9. Grech, M. R., Horberry, T., & Smith, A. Human error in maritime operations: Analyses of accident reports using the Leximancer tool. In *Proceedings of the human factors and ergonomics society annual meeting* 2002 (Vol. 46, No. 19, pp. 1718-1721). Sage CA: Los Angeles, CA: Sage Publications.

10. Grudin, J. *Why CSCW applications fail: Problems in the design and evaluation of organizational interfaces* (pp. 85-93). 1988, Microelectronics and Computer Technology Corporation.

11. Hancock, P. A., & Scallen, S. F. The future of function allocation. *Ergonomics in design*, 1996, *4*(4), 24-29.

12. Hetherington, C., Flin, R., & Mearns, K. Safety in shipping: The human element. *Journal of safety research*, 2006, *37*(4), 401-411.

13. Jirotka, M., & Goguen, J. A. (Eds.). *Requirements engineering: social and technical issues*. 1994, London: Academic Press.

14. *NTSB (2019). Collision between US Navy Destroyer John S McCain and Tanker Alnic MC Singapore Strait, 5 Miles Northeast of Horsburgh Lighthouse". www.ntsb.gov. Retrieved 12 September 2019.*

15. Qureshi, Z. H.. A review of accident modelling approaches for complex socio-technical systems. In *Proceedings of the twelfth Australian workshop on Safety critical systems and software and safety-related programmable systems, 2007, Volume 86* (pp. 47-59). Australian Computer Society, Inc..

16. Rothblum, A. M. (2000). Human error and marine safety. In *National Safety Council Congress and Expo, Orlando, FL 2000,* (No. s 7).

17. Salmon, P. M., Walker, G. H., & Stanton, N. A. Pilot error versus sociotechnical systems failure: a distributed situation awareness analysis of Air France 447. *Theoretical Issues in Ergonomics Science*, 2016, *17*(1), 64-79.

18. Sheridan, T. B. Function allocation: algorithm, alchemy or apostasy?. *International Journal of Human-Computer Studies*, 2000, *52*(2), 203-216.

19. Weick, K. E. *Sensemaking in organizations* (Vol. 3). 1995, Sage.

20. Weigmann, D. A., & Shappell, S. A. (1997). Human factors analysis of postaccident data: Applying theoretical taxonomies of human error. *The International Journal of Aviation Psychology*, *7*(1), 67-81.

TOWARDS AN ECOSYSTEM OF SKILLS IN THE FUTURE MARITIME INDUSTRY

A. Hynnekleiv and **M. Lutzhoft**, Western Norway University of Applied Sciences, Norway
J. V. Earthy, Lloyds Register, UK and Western Norway University of Applied Sciences, Norway

SUMMARY

This research is part of the Human Autonomy Enable (HUMANE) project, addressing human factors in autonomous shipping. The maritime industry is facing a technological evolution that will restructure existing sociotechnical systems. In this highly automated future the role of the human is still unclear. What skills or qualities are needed to interact with advanced technology, both on board and ashore? To explore this issue, and unpack the views of the maritime cluster, we use expert workshops organised in focus group style. The first two workshops invited stakeholders specialising in system safety, cybersecurity and law. Over and above these themes, the discussions also produced a large set of data regarding skills. These data were analysed and used as input values for a third workshop, with maritime education and training experts. This allowed validation of the previously collected data, and a further operationalisation and expansion of the concepts. This paper presents seven sets of skills considered by experts as the most important to make the future sociotechnical systems 'work'.

NOMENCLATURE

HUMANE	Human Autonomy Enable project
STCW	International Convention on Standards of Training, Certification and Watchkeeping for Seafarers
ICT	Information and Communications Technology
KSA	Knowledge, Skill and Ability
HSE	High-Skill Ecosystem
RIA	Robotic, Intelligent, Autonomous
HF	Human Factors
MASS	Maritime Autonomous Surface Ship
AI	Artificial Intelligence

1. INTRODUCTION

The purpose of this study is to contribute to the discussion about skills needed in the future maritime industry. The increase in automation and digitalization introduce many challenging questions regarding the human role in the sociotechnical system [1]: How many crew members will be needed on board? Which crew functions can be moved to shore based centres (if any)? Will the current training defined by STCW be still relevant? A Nautilus study from 2018 shows that a majority of seafarers express anxiety about losing their jobs and being replaced by automation [2]. At the same time, many see a chance for improving safety and working conditions.

In a related paper we present a range of alternatives to teaming humans with intelligent technology and possible applications of these in the maritime domain [3]. The current paper presents what maritime education and training experts believe to be the seven most important skill sets for the future sociotechnical systems.

2. BACKGROUND

2.1 DEFINING SKILLS

In the Human Factors literature, the concept of *skills* is not well defined. It seems that skills belongs to the category of intuitively meaningful concepts, used in scientific discourse as common-sense matter. The domain of Human Factors has been prone to such folk models [4]. The understanding of skills has changed over time, following the socioeconomic settings. The concept of skills gained recognition in the 1970s during discussions on employment policies in response to ICT transformation and growing unemployment. Skill is therefore a concept shaped by social processes [5]. However, definitions for narrow groups of skills can be found in the literature. For example, technical and non-technical skills are often discussed in the maritime human factors field [6, 7].

In this paper we refer to *skills* in a broader sense, comparable to the concept of KSA - Knowledge, Skill and Ability. KSA is an umbrella term covering those human characteristics that can be influenced by training programs [8]. KSA therefore excludes personal characteristics like temper or personality, which are relatively stable over time and cannot be trained. This is the closest definition to the concept of skills as we have used it here. We include traditional skills, knowledge and ability – characteristics that *can* be trained. In the *Results* section, we present skills (in this broader meaning) that are needed in the future sociotechnical system.

2.2 SKILL ECOSYSTEM

The concept of a skill ecosystem has been adapted from biology and applied to economics. A skill ecosystem is defined as *a self-sustaining network of workforce skills and knowledge in an industry or region* [9, p. 5]. There is a range of interdependent factors that allow the development of skills and knowledge: structure of industry, business settings, work culture, job design, education and training system as well as the readiness of individual workers to learn.

Skill ecosystems can be characterised by their demand for low or high skills. Both the present-day and future maritime industry has a demand for high skills. According to Finegold, a high-skill ecosystem (HSE) is *a geographic cluster of organizations (both firms and research institutions) employing staff with advanced, specialized skills in a particular industry and/or technology* [10, p. 61]. HSEs require four elements to be formed and sustained:

- a catalyst – an external trigger which starts the process of skills and knowledge creation
- fuel/nourishment – a factor that maintains the continuous growth
- a supportive host environment – ecological conditions that allow new solutions or ideas to mature
- a high degree of interdependence – interconnections between parts of the system that allow communication and influence [10].

The ecosystem approach can be useful in developing maritime skills on many levels: individual, organisational and industrial. In the maritime context, the term 'ecosystem' is used in connection with MASS and AI. From the marine AI perspective the skill ecosystem is part of/correlated to the marine AI ecosystem which is in turn part of the world AI ecosystem (see *Figure 1*). Maritime skill ecosystem could expand beyond maritime AI.

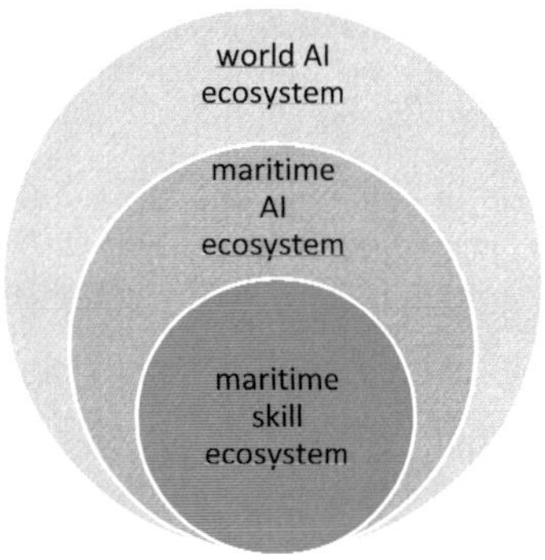

Figure 1: Relationship between world AI ecosystem, maritime AI ecosystem and maritime skill ecosystem (from an AI perspective).

3. METHOD

3.1 PLAN

This paper is one of the outcomes of the HUMANE project. The main method is expert workshops focusing on four different themes (see *Table 1*) connected to autonomous shipping.

3.1 (a) Participants

The workshops involved 59 experts in total (so far) who were invited according to their work experience and current expertise within the workshop's theme. *Table 2.* presents companies and institutions represented by the participants in workshops 1-3.

Table 1: Themes and dates of HUMANE expert workshops.

Workshop	Themes	Date
1.	System safety and cybersecurity	October 2018
2.	Rules, regulations and classification	January 2019
3.	Training and education	November 2019
4.	Work organisation and teamwork	Planned for September 2020

Table 2: Organisations represented in the expert workshops.

CIRM	SINTEF	Inmarsat
Massterly	BW Gas	Wärtsilä
Rolls Royce	Bellona	MTI-NYK
DNV-GL	Lloyd's Register	Wilhelmsen Ship Management
InterManager	ABB	Norcontrol
Kongsberg Maritime	Kongsberg Seatex	Maritime Robotics
F-Secure	RISE Viktoria	EXMAR
Norwegian Maritime Authority	Norwegian Coastal Administration	Swedish Transport Agency
European Maritime Safety Agency	Danish Maritime Authority	International Marine Contractors Association
IMarEST's Maritime Autonomous Surface Ships Special Interest Group	The International Transport Workers' Federation	BIMCO
Safe Marine	Gard	Møkster
InterManager	BW Offshore	SIMAC
BMT Global	SeaBot XR	Western Norway University of Applied Sciences
Norwegian University of Science and Technology	University of South-Eastern Norway	University of Gothenburg
National Maritime College of Ireland	University of Southampton	Åbo Akademi University
Aboa Mare Maritime Academy and Training Center		

3.2 PROCEDURE AND ANALYSIS

The expert workshops were performed in a focus group style. The experts were divided into groups of 5-6 persons. Each group had a moderator with academic background and a note-taker. The discussions were audio-recorded.

3.2 (a) Workshops 1 and 2

During the first two workshops, the participants received a set of prompts related to the themes. These prompts were used to guide the discussion. Audio recordings of the workshops were transcribed, resulting in 400 pages. The transcript was anonymised and coded in NVivo 12 Pro software. Qualitative analysis was performed using thematic grouping and regrouping of coded fragments. The results, apart from the theme-related input, contained numerous comments related to Human Factors, including skills and training. These were synthesised into a list of skills and roles that the workshop participants agreed were needed for the future maritime operators (*Table 3. A-G*).

3.2 (b) Workshop 3.

During the third workshop, the focus was on the skills/roles of an on board operator, identified in the previous two workshops. The original list (see *Table 3*) contained seven roles/skills (A-G) which were then expanded during the third workshop, where the experts added another seven roles/skills (H-O) to the original list. The skills in *Table 3* are explained in the findings section.

The three groups of participants were thereafter instructed to select and discuss the three most important skill/roles from the list, needed to 'make the system work'.

Table 3: Skills / roles used for prioritisation.

	Skills/roles
A	Seamanship
B	Well-trained and multi-skilled
C	IT and cybersecurity literacy
D	Understanding bridge and other electronics
E	Certified mariner (Master's papers)
F	Tool handling
G	Fluent in English
H	Communication
I	Maintenance
J	Digital expert
K	Engineering
L	Awareness of legal issues
M	Leadership of mixed teams
N	Safety awareness
O	Emergency response

4. FINDINGS

This section presents a fuller description of the skills/roles from *Table 3.* based on the expert discussions. The categories are often broad, including many skills, but also knowledge, abilities or attitudes. The categories that were chosen by the experts as the most important for the future sociotechnical system to 'work' are marked with *.

A. *Seamanship** is a common term in the maritime domain, and similar terms are used in many professional areas, such as 'craftsmanship', or 'airmanship'. This is not the traditional seamanship connected to practical deck work, but more of a work ethic of navigators. The consensus on this term is: following the rules of the road, staying out of other ships' way etc.

B. *Well-trained and multi-skilled**, a seemingly cover-all term. It was interpreted by the experts in workshop 3 to mean having the necessary skills and the ability to adapt, as well as knowing how to learn and how to stay current rather than being well-trained in any specific skill.

C. *IT and cybersecurity literacy**. Very often mentioned, as more information technology is introduced, ships become more connected, and the protection seafarers used to have by not being 'connected' is disappearing. These skills are increasingly needed to avoid threats and security breaches. It can be simple things like charging your phone on a piece of bridge equipment or using a private memory stick in ship computers. It could also be much more complex. In workshop 3, this role was complemented with role J – *digital expert*, see below.

D. *Understanding bridge and electronics* – another large role, and an almost impossible task of understanding and being able to operate all electronics and to keep up with development and updates. As an example, during three ship visits performed by the authors in 2018 (a coastal ferry and two supply vessels, built between 2002 and 2012), the number of brands of electronic aids on the ship bridges were counted. The total number of brands were 22, 34 and 40, which gives an indication of the complexity involved in learning and using tools that have been created under a multitude of design paradigms.

E. *Certified mariner (Master's papers)* is mentioned frequently and almost everyone agrees that it is needed. However, it is sometimes mentioned that the 'new' seafarers may not be trained in the same way as 'classical' seafarers.

F. *Tool handling**. This is another catchall phrase. It was used in workshops 1 and 2 to describe the manual work assumed to be needed on board. It includes repairs, maintenance, tweaks, and fixes. The participants of workshop 3 expanded this category to include the handling digital tools, general system knowledge, problem solving skills and a 'can do attitude'.

G. *Fluent in English* – it was commented that those who work in the maritime need a common language, and English was the natural suggestion.

The participants of workshop 3 added the following categories *H-O* before the discussion and selection of the most 'needed to make the system work'.

H. *Communication**. Workshop 3 identified that the ability to work in teams may become the most important skill in the future. This is based on the assumption that the communication will increase, and at the same time more information is needed, and in general work settings will be more dynamic. Surviving and operating in this environment will mean that the traditional self-sufficient seafarer of today is the opposite of what we will need in the future. How much of this is related to personality is an issue for investigation.

I. *Maintenance* – mentioned before as part of *tool handling*, but also created as a separate category. The focus was set especially on planning the maintenance procedures.

J. *Digital expert* was added because the participants of workshop 3 considered the need for a broader concept than *IT literate*; a strategic level skill, allowing operational supervision. One participant said that the concept also included overall logistics.

K. *Engineering* was suggested at workshop 3, as none of the earlier concepts covered it. One of the participants suggested that the trends in available expertise, propulsion technologies and supplier capability are moving towards systems that are more simple to manage and easy to maintain. The need for engineers on board is going to decrease. On the other hand, other new specialisations will be needed on board (e.g. configuration management, IT, data management).

L. *Awareness of legal issues*. It was assumed to include also considerations of future responsibility and authority.

M. *Leadership of mixed teams* includes teams with varying nationalities but also human-machine teams. The future scenarios will involve the need for interim and short-term cooperation. Therefore, beside team formation, the focus should also be on 'joint activity' or 'flash teams'.

N. *Safety awareness** – including a new set of risks related to automation and other new technologies. This relates to being able to build and maintain a safety culture, promoting good practice and operational discipline ('doing the right thing every time').

O. *Emergency response** – similar to today, but with the prospect of having fewer people to assist. Emergency response requires system knowledge and understanding. The focus was on having the ability to gather critical data, analyse it and create an action plan. A need to include communication and management skills was also pointed out.

The seven most important skills (see *Figure 2*) are strongly related to an ability to learn and relearn, adapt and manage new situations. This paints a picture of the future maritime employee as an adaptive and resilient individual. Our results correspond with future skills forecasting projects: World Economic Forum [12] and IFTF [13] – two wide-scoped reports that stress the importance of adaptive skills, ability to learn and understanding the technology.

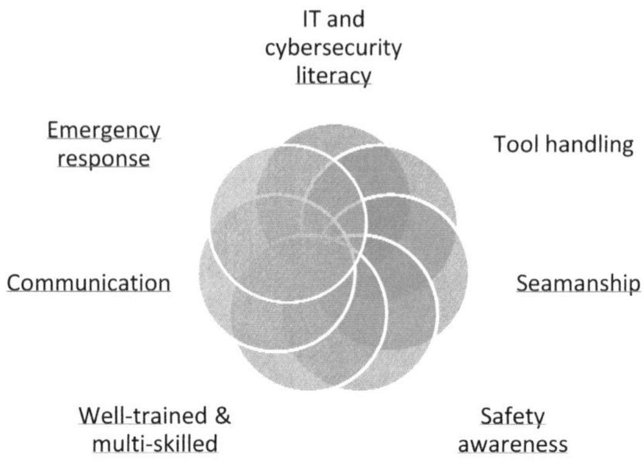

Figure 2: Skills chosen by the experts as most important for the future sociotechnical system to 'work'.

7. DISCUSSION

During the workshop, the groups focused on the competence needs for high levels of automation, and also identified an organisational structure that would support this competence. This supporting structure is not the one currently in use in the maritime industry and indeed there was some evidence that the current industry structure would be inimical to the development and maintenance of this competence.

It was concluded that if the maritime industry keeps its current structure, it might not be able to make safe, effective and full use of high automation. On the other hand, our guidance on how to adopt high automation may need to include directions on structural change for the industry. For example, manning is currently outsourced. How can the necessary competence be developed, or teams be formed, in this environment?

There was also a long discussion on screening, monitoring, enhancing and/or augmenting humans, and not a lot of discussion on quality of life, welfare or partnership with technology. Seafarer well-being is challenged by work-related factors, often leading to depression. Suicide statistics indicate that the mental health of seafarers continues to be very poor [11].

To connect back to the ecosystem model, we need four elements to support a maritime skill ecosystem:

- A trigger, or many. The external stimulus which starts the process of skills and knowledge creation should appear early in the process, not after the technology has been designed. In the present case, the trigger is the push for autonomous ships and the skill creation is long overdue.
- Fuel/nourishment – a factor that maintains the continuous growth – could be collaborations between industry and academia, and associated

support for such collaboration. Also, Industry 4.0 can be regarded as nourishment, as shipping needs to become more connected to data-centric business.

- A supportive host environment – ecological conditions that allow new solutions or ideas to mature. This could be implemented by authorities and governments, for example to achieve the same or higher levels of safety. It can also include infrastructure.
- A high degree of interdependence – interconnections between parts of the system to allow communication and influence – this could be knowledge sharing networks, including international exchange [10].

8. CONCLUSIONS

The result of the exercise reveals several skills/roles that are considered important for future maritime personnel. Because the clusters overlap, we are not presenting absolute *numbers*, but rather unpacking the categories of skills.

According to the participants, there was no need for a clear division of skills between ship and shore staff. Furthermore, the result is a mix of old and new, relatively generic skills, and some that can be considered traits rather than conventional academic subjects. The industry should be looking for confident and adaptive individuals, who can manage emergencies, communicate seamlessly, are practical to some degree and well schooled in IT and software. A strong emphasis is placed on knowing how to learn and relearn, adapt and manage new situations.

We also see that for a healthy ecosystem, there needs to be high-level buy-in and support for collaboration, which points back to the finding that the structure of industry needs to change.

9. REFERENCES

1. JOHNS, M., 'Seafarers and digital disruption', *Hamburg School of Business Administration*, 2018.
2. NAUTILUS FEDERATION, 'Future proofed? What maritime professionals think about autonomous shipping', *Available: https://nautilusfederation.org/wp-content/uploads/2018/02/report_auto_8.2.18.compressed.pdf,* 2018.
3. HYNNEKLEIV, A., LUTZHOFT, M., EARTHY, J. V., 'Siri, sail the ship! - Exploring human-RIA relationships in the maritime domain', *Ergoship*, 2019.
4. DEKKER S., HOLLNAGEL, E., 'Human factors and folk models.' *Cognition, Technology & Work*, 2004.
5. RIGBY, M., SANCHIS, E., 'The concept of skill and its social construction.' *European journal of vocational training*, 2006.
6. HORBERRY, T., GRECH, M., KOESTER, T., 'Human factors in the maritime domain', *CRC press*, 2008.
7. BARNETT, M., GATFIELD, D., PEKCAN, C., 'Non-technical skills: the vital ingredient in world maritime technology', *2nd World maritime technology conference*, 2006.
8. STEVENS, M. J., CAMPION, M. A., 'The knowledge, skill, and ability requirements for teamwork: Implications for human resource management', *Journal of management*, 1994.
9. FINEGOLD, D., 'Creating self-sustaining, high-skill ecosystems', *Oxford review of economic policy*, 1999.
10. WINDSOR, K., ALCORSO, C., 'Skills in context: A guide to the skill ecosystem approach to workforce development', *Available: https://www.voced.edu.au/content/ngv:12460,* 2008.
11. IVERSEN, R. T. B., 'The mental health of seafarers', *International maritime health*, 2012.
12. WORLD ECONOMIC FORUM, 'The future of jobs report 2018', *Available: https://www.voced.edu.au/content/ngv:80780*, 2018.
13. DAVIES, A., FIDLER, D., GORBIS, M., 'Future work skills 2020. Institute for the Future for University of Phoenix Research Institute', 2011.

10. AUTHORS BIOGRAPHY

Agnieszka Hynnekleiv is a PhD student in Maritime Human Factors at Western Norway University of Applied Sciences and a team member of HUMANE project. Her doctoral research investigates future skills for the maritime personnel. She holds a master's degree in psychology and her previous experience includes designing training programmes for different occupational groups.

Margareta Lutzhoft is Professor of Human Factors at Western Norway University of Applied Sciences and the leader of HUMANE. She has a background as a master mariner, a PhD in Human-Machine Interaction and MSc in computer science. Her main interest is design for safety and the effects of new technology and automation on humans.

Jonathan Earthy is technical authority for HF in Lloyd's Register, including the R&D programme and the introduction of Ergonomics into Class. His technical speciality is the assurance of the quality of Human Factors Integration and HCD, with special reference to advanced ICT. He has participated in standards development since the mid-1980s and is convenor of ISO TC159/SC4/WG6 HCD for interactive systems.

MOVING HAVENS
AN APPLICATION OF THE E-NAVIGATION SERVICE ROUTE EXCHANGE

T. Porathe, Norwegian University of Science and Technology, Norway

SUMMARY

e-Navigation is a concept launched by the IMO in 2006. The point is to share digital data to the benefit of safety, efficiency and the environment. Every ship must make a voyage plan before leaving berth. Sharing such voyage plans between ships and between ships and shore is called route exchange and such a feature would allow alarms to be triggered both onboard and ashore if a ship deviate from its planned route, or if two ships plan to be at the same place at the same time. To visualise correspondence or lack of correspondence between the planed position and the real position the concept of Moving Havens is suggested. It is a visualisation tool designed to intuitively show if a ship falls out of its safety checked position. In case of route exchange and Ship Traffic Management it will facilitate for land-based operators or automation to distinguish dangerous situations. The concept is presented in this paper.

1. MAKASSAR HIGHWAY

In the early morning hours of 23 July 2018, the 139 meters long car carrier MAKASSAR HIGHWAY grounded on the Swedish east coast (see Figure 1). There were no fatalities, but the ship was badly damaged and had to be scrapped. The vessel had deviated from its planned and safety checked track in order to go closer to the coast to ensure cell phone and Internet coverage. At the time of the grounding the vessel was steaming at economy speed, somewhat less than 14 knots, in smooth sea and good visability on a straight course in among the small skerries, rocks and shallows on a remote part of the Swedish archipelago south of Stockholm (see Figure 2).

Figure 1: MAKASSAR HIGHWAY grounded on the Swedish east coast on 23 July 2018. [1]

As the ship was approaching the coast at 7 o'clock in the morning (local lime), the first officer who had the watch was either asleep or at the back of the bridge doing administrative work. The accident investigation failed to establish exactly what he was doing, only that he was not monitoring the progress of the ship. He had a blood alcohol content at the time of the grounding estimated to 1.15 permille, according to the police report [1]. The lookout was absent from the bridge on his fire watch round when the vessel grounded.

MAKASSAR HIGHWAY had the morning before left the Kiel canal for the final leg to Sodertalje, south of Stockholm, with 1325 cars and a crew of 20 onboard. The vessel had, in accordance with Chapter V, rule 34 of SOLAS [2] and Section A-VIII/2 of STCW [3], the stipulated voyage plan programmed into the Electronic Chart and Display Information System (ECDIS). In order to monitor the vessel's progress along this planned track, the track would have had to be safety checked for under-keel clearance in order to make sure the planned track was not passing into shallow water.

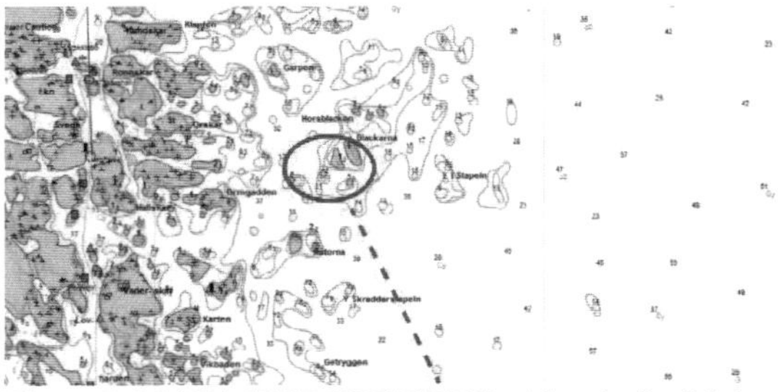

Figure 2: MAKASSAR HIGHWAY with a draft of 6.6 m grounded SE of the small port of Flatvarp on the Swedish east coast at 7:21, local time, the 23 July 2018. Derived from positions in [1].

However, during the afternoon on the day before, the vessel left the planned track and followed the coast line in the Bay of Hano (see Deviations from the track, in Figure 3). The aim was to reach the mobile telephone network and thereby having a chance to connect to the Internet at a reasonable cost in order to call family, etc. This course adjustment was accepted by the master. However, the vessel's voyage plan was not updated (which it should, according to the regulation referred to above).

During the evening of the 22nd another such deviation was made as MAKASSAR HIGHWAY went closer to the coast of the island of Oland, and later made the fatal coarse change that lead to the grounding. In Figure 3 the safety-checked voyage plan is showed in red, and the actual track the ship took is shown in blue.

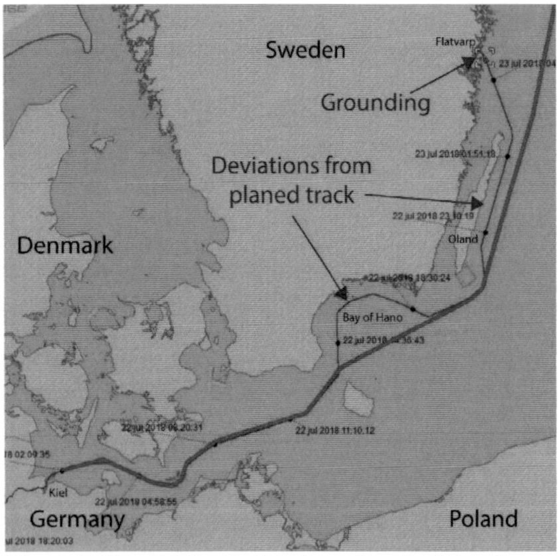

Figure 3: The red line is the planned and validated route plan for MAKASSAR HIGHWAY's voyage between Kiel, Germany, towards Sodertalje, Sweden. The blue line is the actual AIS track sailed. Times here are in UTC, i.e. Swedish summer time +2H. Adapted from [1].

Swedish territorial waters in monitored by three different authorities: the Swedish Maritime Administration's coastal Vessel Traffic Service (VTS), the Coast Guard and the Defence. None of these entities raised any alarm when Makassar Highway deviated from its planned track (which is natural as voyage plans are presently not shared with external bodies). However, the VTS in Sodertalje notified the Swedish Jointed Rescue Coordination Centre (JRCC) in Gothenburg 12 minutes after the grounding that MAKASSAR HIGHWAY "had been observed being still outside of Vastervik" [1]. At around 07:45, the JRCC contacted the vessel via VHF radio. The person who answered on board stated that: "...since seven thirty we are aground..." [1].

During 2018 and 2019 two more similar groundings, with incapacitated watch officers, have occurred in Swedish waters.

1.2 OFF-TRACK ALARMS

If a watch officer falls asleep or is distracted and a ship deviates from its planed sailing route, should there not be an alarm? In the case of MAKASSAR HIGHWAY there was indeed a Bridge Navigation Watch Alarm System (BNWAS), much like a wake-up clock which sounded every 12 minutes if there was no activity on the bridge. The timer then had to be reset for another 12-minute period. This system was turned off.

However, the ECDIS also includes some automatic features designed to assist the watch officer for precisely this reason. One of them are based on the existence of a voyage plan.

The IMO regulation says: "On the basis of the fullest possible appraisal, a detailed voyage or passage plan should be prepared which should cover the entire voyage or passage from berth to berth," [4].

A voyage plan in an ECDIS is made by clicking out way points along the intended route. Some attributes need to be added to this track-line, one of them is *cross-track distance (XTD)*. Own vessels *safety depth* also needs to be set. The safety depth is the needed water depth considering draught, squat and a safety margin. When this is done the route can be automatically checked for under-keel clearance within a corridor limited by the port and starboard XTD (see Figure 4).

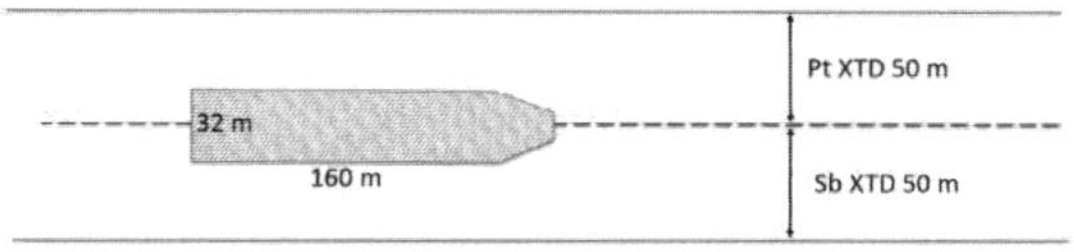

Figure 4: Cross-track distance (XTD), is set to a number of meters on the port and starboard side of the dotted track line. The corridor can then be automatically checked in ECDIS for under-keep clearance. An alarm can also be set to sound if the vessel leaves the corridor.

"When Cross Track Distances are properly set to each leg of a voyage plan then route checking assists in checking for potential obstructions, dangers and insufficient depths" [5], i.e. warnings are presented if a track line for instance passes over land, or over areas with a water depth less than the set Safety Depth.

A vessels autopilot can even be set to "path-following" mode, whereby the ship automatically can pass along a track. On one Hurtigruten ship (Norwegian coastal express) which I visited on the bridge, the captain said that they mostly sailed using track-following mode, with an off-track alarm set to 50 meters port and starboard. Manual steering was only used in very narrow passages and when going to and from mooring.

All ECDIS also have an off-track alarm for leaving the planned track. "It should be possible for the mariner to specify a cross track limit of deviation from the planned route at which an automatic off-track alarm should be activated" [6].

There is no mentioning of an off-track alarm in the accident report. And the captain states that he authorised a deviation from the planed track in the Bay of Hano the night before the grounding. This deviation was done without changing the voyage plan. However, he denies authorising the second deviation that eventually led to the grounding. It is therefore follows that this off-track alarm must have been turned off.

Many ECDIS also have a safety feature called "look-ahead sector" (or similarly). This feature warns if the water depth within a set distance in front to the vessel is less than the set Safety Depth. This feature must also have been turned off (as it often is due to many false alarms in narrow waterways). But why was it not turned on as MAKKASAR HIGHWAY had been transiting open waters during the last 24 hours?

Had it been on, had the voyage plan been updated to reflect the new deviations and had the off-track alarm been on, and then the watch officer then fell asleep or became distracted, an alarm would call for his attention if he missed his turn. If he should he not react, such alarms can in many ships be set to promulgate to another officer onboard, the captain or to sound in the general crew's quarters. It is reasonable to believe that this could have prevented the accident.

One may wonder why such alarms can be turned off? On the other hand, experiences from interviewing bridge officers during many years show that the "alarm hysteria" on the bridge is a big problem. One may also wonder if it would not be useful if such an off-track alarm also was visible for the coastal VTS? This last question is addressed by *e-Navigation*.

2. E-NAVIGTION

The basic technologies of electronic navigation (apart from radio and radar) came with the Global Positioning System (GPS) and electronic chart plotters in the 1980's and 90's. At about the same time in the maritime domain the Swedish pilot Benny Pettersson started experimenting with transmitting GPS positions through the Ship Position Exchange System (SPEX), later to be named Automatic Identification System (AIS). AIS became mandatory by the IMO in 2002.

In 2006 IMO approved a proposal by Japan, Marshall Islands, Netherlands, Norway, Singapore, United Kingdom and United States to develop an "e-Navigation strategy". The objective of the proposal was to "develop a broad strategic vision for incorporating the use of new technologies in a structured way and ensuring that their use is compliant with various navigational communication technologies and services that are already available, with the aim of developing an overarching accurate, secure and cost-effective system with the potential to provide global coverage for ships of all sizes" [7].

The point was to share digital data to the benefit of safety, efficiency and protection of the environment. As an example: if ships where to share voyage plans between between themselves and shore (such as VTS), "single points of failure" could be caught, such as a ship straying off its course, whether deliberate or because a watch officer was making an error or falling asleep. In many research projects since, possible e-Navigation features has been investigated. The feasibility to share routes has been called "route exchange" and could potentially allow alarms to be triggered, first onboard and later ashore, if a ship deviated from its planned route. Addressing precisely the accident above. Route exchange in different forms has been researched by e.g. [8, 9]. However, there are also cases where alarm triggered by off-track, or shallow water cannot prevent accidents. Such cases are collisions and an example will follow.

3. THE FU SHAN HAI-GDYNIA COLLISION

On a beautiful May day in 2003 the 225-meter-long Chinese bulk carrier FU SHAN HAI collided with the 100-meter-long Cyprus registered container ship GDYNIA about 3 nautical miles north of the island of Bornholm in the southern Baltic Sea. The ships had been on straight courses and in sight of each other for hours before the collision. According to Rule 15 of the International Regulations for Preventing Collisions at Sea 1972 [10] FU SHAN HAI was the stand-on vessel with right of way (see Figure 5).

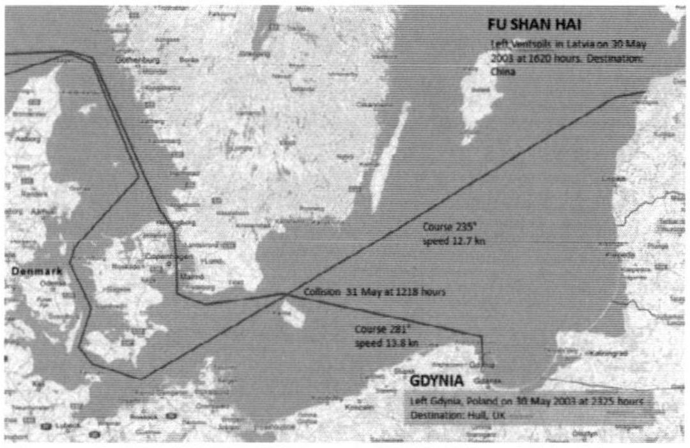

Figure 5: FU SHAN HAI and GDYNIA collided in open sea and good visibility 31 May 2003 [Adapted from 11].

The Chinese vessel was by the rules compelled to keep her course and speed. GDYNIA, on the other hand, being the give-way vessel, should by the same rules either turn starboard and pass behind FU SHAN HAI, or slow down and let FU SHAN HAI pass in front of her. Neither was done. The reasons are unclear. There was an exchange of watches on the GDYNIA just before the collision. But it looks as the Chinese vessel was not at all observed from the GDYIA until it was too late. The subsequent collision ripped open a hole in the forward part of FU SHAN HAI which sank (see Figure 6). The crew of 27 were all rescued. [11].

Figure 6: FU SHAN HAI sinking after the collision, [11].

FU SHAN HAI had departed from Ventspils in Latvia on 30 May 2003 at 16:20 local time on a voyage to China. GDYNIA had departed from port of Gdynia in Poland the same evening at 23:25 on a voyage to Hull, England. This collision serves as an example of the about 10-15 vessels every year that are totally lost after a collision at sea. Although there is a lot to be said about the actions of the crews in relation to COLREGS, this paper will look if features made available by the e-Navigation development could be used to prevent this type of accidents.

3.1 TIME COORDINATIED VOYAGE PLANS

Both FU SHAN HAI and GDYNIA had onboard the required voyage plans, and were sailing in accordance with these plans, so no off-track alarm could warn then of the upcoming encounter ahead of time. But, if we assume that these two voyage plans had been transmitted to some sort of central coordination mechanism and superimposed on top of each other, then a warning might have been issued.

The IMO guidelines for voyage planning [4] also mentions the need to determine the estimated time of arrival (ETA) both at the destination and "at critical points for tide heights and flow". Until recently standard format for route exchange has not included speed settings and ETA at waypoints. However, in the MONALISA projects [12, 13] the author took part in work which resulted in the so called RTZ-format for route exchange, approved by the International Electrotechnical Commission in 2015 [14]. This format allows precisely that, and would allow time coordinated voyage plans to be compared for the existence of "loss of separation", i.e. if two ships planned to be in the same place at the same time.

In reality, a prediction made maybe days ahead will be inexact as speed might change due to weather, engine performance, sea state and many other parameters. But if the onboard ECDIS keeps comparing present and planned positions, safety checks and shares a constantly updated voyage plan with the central coordination mechanism the prediction will become better and better as the time to the predicted moment decreases. In the case of FU SHAN HAI and GDYNIA such a system could have alerted the bridge crew on both ships about the predicted close quarter's situation and might have recommended a minor speed change on one or both ships. Furthermore, information about the upcoming situation would have been available in the central coordination centre (maybe a "Baltic VTS") where yet another pair of eyes would be alerted and could intervene. However, to simplify for the navigator onboard and the operator in a VTS a visualisation tool is needed. This tool is called Moving Haven.

4. MOVIG HAVENS

When submarines operate together in groups they cannot see or hear each other under water. To avoid collisions, they use coordinated voyage plans in three dimensions. Each submarine is designated to a "moving haven" visualized as a cube, moving in a 3D nautical chart. The own submarine's motions are tracked by an advanced inertial navigation system and the navigator's task is to keep the submarine within its own designated box. The other submarines' positions are visible through their havens, given that they all stay in their boxes [14].

The same principle can be applied to surface vessels. By visualizing the planned position of ship along its voyage plan with a box, a Moving Haven, the navigation officer onboard can have a quick and intuitive confirmation that he is on track and on time. If he or she strays out of the box there could be a warning.

4.1 THE WIDTH – CROSS TRACK DISTANCE

Cross-track distance (XTD) was discussed above in section 1.2. By attributing an XTD on each side of the track in the voyage plan, a corridor of safety checked water can be created for the ship. By tailoring this XTD for each leg, smaller in confined waters and larger in open sea, a dynamic precision in navigation can be acquired. In some cases, the XTD can be different on the port and starboard side, e.g. when the track passes close to a buoy. The corridor created by the port and starboard XTD can for each leg be used as the width of the Moving Haven. See Figure 7.

4.2 THE LENGTH – TEMPORAL PRECISION

The length of the Moving Haven has to do with the needed temporal precision and effective space management in a traffic coordinated system.

The length of a Moving Haven set for a precision of one hour with a ship moving at 15 knots would be 15 nautical miles long. This is not a "box", more like a long "snake". In a time-coordinated ship traffic management system where the Moving Haven was used as a "safe haven" which would be exclusive for only one ship, this would mean that the ship would block this whole area. This would be inefficient in congested waters. Some other concepts need to be considered.

Just in time arrival is a logistic concept used to make traffic flows more efficient and decrease greenhouse gas emissions. Traditionally ships would steam along their rout on their standard speed and arrive to their destination early to be able to anchor and issue a *notice of readiness*. Depending on the charter agreement they could then collect *demurrage* while waiting for the port to be ready to take them in. However, if the readiness of the port and the arrival of the ship is synchronised the ship might be able to *slow steam* to its destination and arrive just in time. And because fuel consumption (and accompanying emissions) depends exponentially on the speed, large savings can be made by slowing down.

A port is ready to take in a vessel when all assets are in place: the tide is right, pilots are ready, the required length of pier side is free, tugboats and linesmen are in place, etc. There might even be a booked time for a lock passage to reach the berth. In any of these cases a time precision for the estimated time of arrival (ETA) could be less than 5 minutes. An assumed precision of 1 minute for a 160 metres long ship moving in 15 knots is illustrated in Figure 6. That Moving Haven is 2.5 cables long (463 meters) and 100 meters wide, assuming a XTD of 50 meters port and starboard. There should be no problem to stay in such a box in nice weather for an autopilot in track-following-mode and a good *speed pilot* (autopilot for speed). Heavy wind, waves and currents would of course require different dimensions.

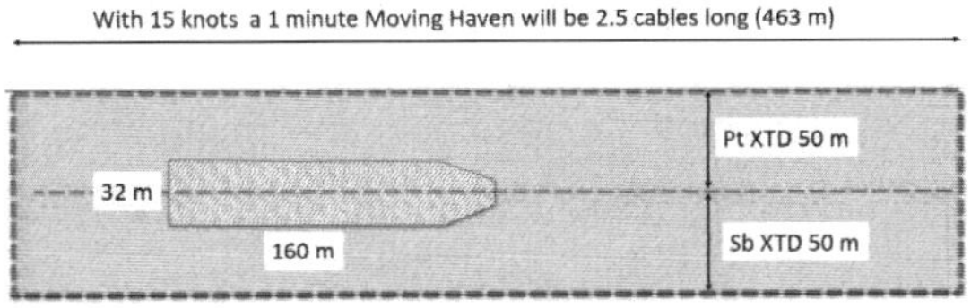

Figure 7: A 1-minute Moving Haven for a 160 meters long vessel at 15 knots will be 463 meters long. The width will be 100 meters with a XTD of 50 meters on either side of the track line. On the ECDIS the colour is green because own ship is in the haven, "on track and on time".

The time-precision could change dynamically during the voyage. Less stringent at the start and more precise closer to destination. If a ship slips out of its Moving Haven an alarm would be triggered. The ship should then either get back into the box, update the track or recalculate ETAs for the voyage.

4.3 ALARMS

Alarms could, as mentioned above, be given to the watch officer for the case the ship gets off track or loses its time slot. This could be done first visually with colours, as suggested in Figures 6, 7 and 8. If there is no response they could promulgate with alarms, first within the ship, and finally, in a STM regime, sent to a coordinating mechanism, e.g. a VTS, if one is in place for the waters.

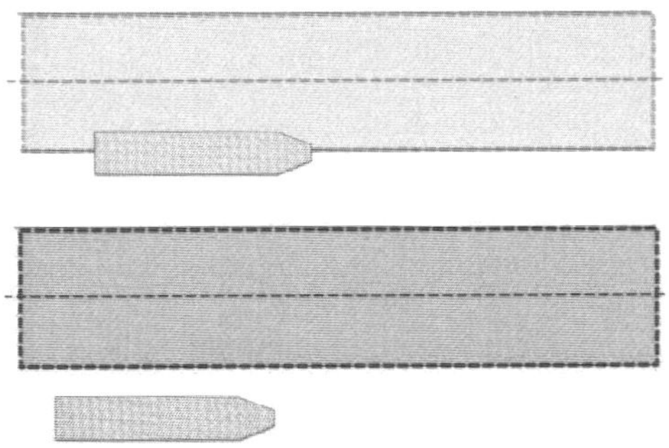

Figure 8: If a ship starts to slip out of its Moving Haven a colour warning could first alert the watch officer.

In case such a coordinating mechanism is in place and routes are shared the scenario for the example accident with FU SHAN HAI and GDYNIA, referred to above, could look like Figure 9, or 10.

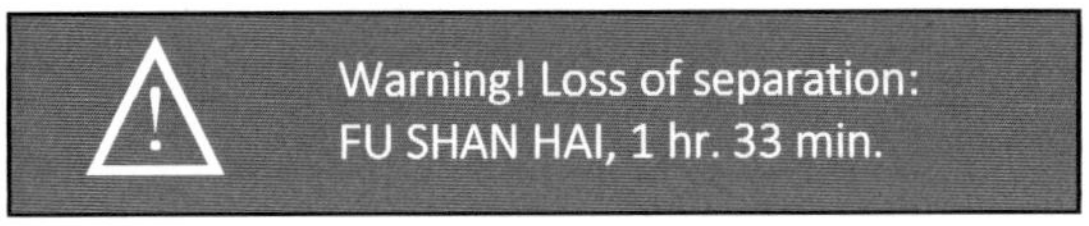

Figure 9: A central coordination system receiving voyage plans from all ships reacting to is two ships were to approach a point where they would be at the same place at the same time.

In the MONALISA project technical tests with the Moving Haven in the ECDIS-like "e-Navigation Prototype Display" were made onboard a Korean training vessel in the southern part of the Korean Republic. Figure 10 shows a screen dump from these tests. For safety reasons the ship was not allowed to navigate using the Moving Haven why the ship on this chart is outside and above the "box" in the red circle. On the "conning display", bottom right, the Haven is coloured red to warn for this fact [8].

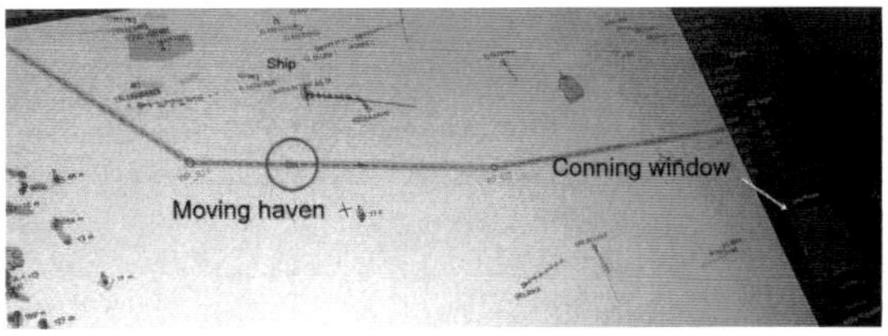

Figure 10: A screen photo from an ECDIS-like prototype display system using a Moving Haven in the sea outside, Wando in South Korea during a test in 2014. The own ship is outside ad north of the Moving Haven [8]. Photo by the author.

5. CONCLUSIONS

e-Navigation is about to deliver new infrastructure for sharing route information. This makes it possible to allow ship traffic management systems to survey and intervene in some type of accidents where bridge officers onboard have lost situation awareness (single point of failure).

Following a voyage plan for a set destination and ETA means turning up at specified waypoints at specified times, i.e. following a scheduled "virtual position" along a track on the ECDIS. This "virtual position" today often recides in the mind of the watch officer which makes it difficult to share with the outside world. Visualising this "virtual position" with a Moving Haven would make a deviation from the plan obvious for the outside world which then has a bigger chase of intervening.

While the AIS system will visualize the position of the real ship in the ECDIS, the Moving Haven will visualize the planned (and possibly) coordinated position where the ship is supposed to be. If all ships stay in their Moving Havens, close quarter's situations and groundings can be avoided. At least in theory. In reality there are also fishing boats and leisure crafts that are not part of the system. But id at least all larger SOLAS vessels are included safety can be increased.

6. ACKNOWLEDGEMENTS

The research has partly been conducted within the SESAME Solution 2 (Secure, Efficient and Safe maritime traffic Management in the Straits of Malacca and Singapore) project, and partly within the IMAT (Integrated Maritime Autonomous Transport systems) project, both funded by the Norwegian Research Council, which is hereby gratefully acknowledged.

7. REFERENCES

1. SWEDISH ACCIDENT INVESTIGATION AUTHORITY, 'Final report RS 2019:04e, MAKASSAR HIGHWAY – grounding with subsequent oil spill off Västervik, Kalmar County, on 23 July 2018, File no. S-148/18, 9 July 2019.

2. IMO, International Maritime Organization, 'SOLAS – International Convention for the Safety of Life at Sea', 1974, as amended.

3. IMO, International Maritime Organization, 'STCW – Standards of Training, Certification and Watchkeeping for seafarers', 2010.

4. IMO, International Maritime Organization, 'Guidelines for Voyage Planning', Resolution A.893(21), adopted on 25 November 1999.

5. THE LONDON P&I CLUB, 'Is your ECDIS contributing to safe navigation or introducing risk?' LP Focus (7), June 2018.

6. IMO, International Maritime Organization, 'Revised Performance Standards ECDIS', Resolution MSC.232(82), 2006.

7. IMO, International Maritime Organization, 'Maritime Safety Committee, Preliminary assessment of proposals for new work programme items', MSC 81/23/10, 2006.

8. PORATHE, T., BORUP, O., JEONG, J.S., PARK, J., BRODJE, A. & ANDERSEN CAMRE, D., 'Ship traffic management route exchange: acceptance in Korea and Sweden, a cross cultural study', Proceedings of the International Symposium Information on Ships, ISIS 2014, Hamburg, Germany, 2014.

9. PORATHE, T., BRODJE, A., WEBER, R., CAMRE, D. & BORUP, O., 'Supporting Situation Awareness on the bridge: testing route exchange in a practical e-Navigation study', in A. WEINTRIT, & T. NEUMANN (Eds.) 'Information, Communication and Environment: Marine Navigation and Safety of Sea Transportation', CRC Press, 85-92, 2015

10. IMO, the International Maritime Organization. (1972). Convention on the International Regulations for Preventing Collisions at Sea, 1972, (COLREGs). London: IMO.

11. DMA, Danish Maritime Authority, 'Collision between Chinese bulk carrier FU SHAN HAI and Cypriot container vessel GDYNIA May 31, 2003', Division for Investigation of Maritime Accidents, File 199935831, August 7, 2003.

12. MONALISA project, Swedish Maritime Administration, http://sjofartsverket.se/monalisa [acc. 2020.01.13]

13. MONALISA 2 project, Swedish Maritime Administration, http://monalisaproject.eu/ [acc. 2020.01.13]

14. IEC, International Electrotechnical Commission, 61174, https://webstore.iec.ch/publication/23128 [acc. 2020.01.13]

15. Porathe, pers. com.

9. AUTHORS BIOGRAPHY

Dr Thomas Porathe holds the current position of professor of interaction design at the Dep. of Design at NTNU, the Norwegian University of Science and Technology in Trondheim. He has for many years worked with design of maritime information systems, e-Navigation, Human Factors and MASS. He is contributing to the work in the IALA's ARM committee.

HUMAN FACTORS RELATED TO AUTONOMOUS SHIP OPERATIONS

E. Aa. Dahle, Dr. Ing., Norway
K. Bentzrod, Hoegh LNG, Norway
N. H. Lund, Master Mariner and Maritime Government Pilot (Ret.), Norway

SUMMARY

Test operation of autonomous small cargo ships, in short, restricted inshore lanes is presently undertaken in Norway. The goal is to eliminate onboard crew and associated cost components, and human failures committed onboard.
The running of the ship will to a large extent be taken over by automation, and the ship will be operated by remote control from shore. Hence, human errors committed by remote operation ashore may be restricted to ship and cargo damage. A notable exception is passenger ships, where evacuation must be assisted by a trained crew. Also, the danger classification of the cargo may restrict the degree of remote operation. For all ships, the most prominent human errors committed by the crew are related to navigation. Here, powered grounding dominates. It is considered that the main reason is a lack of structure and control in the operation of ships, i.e. leaving too much freedom to the ship officers. There is a reluctance to restrict this traditional freedom to act in accordance with a personal judgement. However, on coastal areas with dense traffic, as well as in sea areas with e.g. oil installations and wind farms, traffic regulations schemes and VTS surveillance may be necessary. Combination of autonomous ships and ships with traditional navigation in a fairway is considered to increase the accident frequency.

NOMENCLATURE

- AIS. Automatic Identification System. AIS. The system is mandatory for ships above 45m in length, or with tonnage above 300GRT. The system submits information to other ships and also to VTS. The info consists of i.a. position, course and speed over the ground
- AI. Articial Intelligens, a computer system programmed for intelligent learning
- AToN. Aids to Navigation. Physical means, or virtual by AIS
- BNWAS. Bridge Navigation Watch Alarm System. The system is mandatory for ships above 250BRT. If the OOW does not respond, alarm is submitted to a manned station onboard, i.e. a "dead man button", often based upon immobility on the bridge. Also, leisure crafts may acquire a relatively simple AIS. A corresponding system must be installed in the future VTS
- COLREGS. The International Regulations for Preventing Collisions at Sea. International rules of conduct for commercial ships. Includes rules for lights and signals, and TSS
- ECDIS. Electronic Chart and Display System. The system may be fully implemented from 2018. The screen-based system will display relevant electronic charts with selected data from radar, satellite systems (up to 4 are available) Guidelines in SOLAS
- E-navigation. A ship with ECDIS must i.a. submit its sailing plan for approval to the relevant VTS
- IMO. International Maritime Organization
- OOW. Officer On Watch (on a ship or in a VTS ashore)
- SOLAS. The International Convention for the Safety of Life at Sea. An international maritime treaty obliging signatory states to minimum standards in construction, equipment and operation
- STCW. The International Convention on Standards of Training, Certification and Watch keeping for Seafarers. Specifying qualification requirement for all on board maritime personnel. Guidelines in SOLAS
- VHF. Very High Frequency radio, used on VTS, and on most ships and vessels. For coastal areas only, due to limited range
- TSS. Traffic Separation Scheme. Such systems will define a sector, normally one for each direction that shall be followed under normal condition. Smaller ships may be assigned a limited, outlaying sector. Guidelines in COLREGS
- VTS Vessel Traffic Service, similar to air traffic control for aircraft and road administrations. In Norway there are 5 VTS, which survey particularly sensitive sea areas along the coast. VTS use i.a. VHF, AIS and radar. Guidelines in SOLAS

1. INTRODUCTION

Autonomous ships can be divided into groups;

1. Remote operated ships are operated by humans on shore.
2. An automated ship can follow a preset course line automatically, supervised by humans onboard. Otherwise traditional navigation

3. A fully autonomous ship is run by artificial intelligence (AI), without human intervention

In this paper, only remote operated ships and automated ships are covered, and a comparison is made between the two.
The reason is that the authors do not believe that it feasible to run a fully autonomous ship within the present regimes at sea. Inter alia, passenger ships evacuation must be assisted by a trained crew Also, the danger classification of the cargo may restrict the degree of remote operation. Furthermore, it is not considered feasible to leave considerations inherent in COLREGS Regulations 17 and 19 to AI.

For all ships, the most prominent human errors committed by the crew are related to navigation. It is considered that the main reason is lack of control in the operation of ships, i.e. leaving too much freedom to the ship officers. Furthermore, in coastal areas with dense traffic, as well as in sea areas with e.g. oil installations and wind farms, strict traffic separation schemes and VTS surveillance and control may be necessary. Although under study by IMO, such
restrictions are not likely to be accepted. There is a reluctance to restrict the traditional freedom to act in accordance with a personal judgement. An attempt to replace this freedom with AI is considered as futile. A mix of fully autonomous ship with traditional ships in a common fairway, is considered to increase the accident frequency. This assessment is only to some extent related to powered grounding. The negative attitude of the authors is mostly related to collisions, for which AI is considered as not suited within the present context of COLREGS. Collisions are not discussed in this paper, only powered grounding. Only short, coastal fairways strictly reserved for fully autonomous ships at moderate speed are considered feasible for AI implementation.
From the point of view of risk analyses, risk reduction could be attained by reduction of frequency of human failures, and/or by reduction of grounding consequences. The latter is not a subject of this paper.

2. AUTONOMOUS OPERATIONS IN GENERAL

At present, fully autonomous cars are tested out. The experience is mixed, and the first fatality to a pedestrian has occurred. The reason seems to be twofold; imperfect registration of traffic signals, and lack of ability to register the mental state of pedestrians (distracted, old, infirm, blind). This is comparable to a "common fairway", i.e. pedestrians should be separated from wheel traffic lanes. Separation was suggested by Leonardo de Vinci more than 400 years ago and is a sound principle even today.
Railway traffic is ideally suited for autonomous implementation. For decades, autonomous train have been running to and from airports, and between air terminals. The most important risk is eliminated by automatic stop if doors are obstructed, or if the trail is blocked or damaged.
Air traffic is a so new that strict traffic separation schemes and traffic surveillance and control measures have been readily accepted, leaving very little freedom of choice to the pilot.
A corresponding system can in theory be the solution for autonomous ship traffic.
Experimental running of models of autonomous small cargo ships, in short, restricted inshore lanes is presently undertaken in several countries.
In [2] an example is presented on the development of a model of an autonomous inshore electric container ship, the Norwegian Yara Birkeland. A full scale ship is expected to be completed in 2020 and is to be fully autonomous by 2022. The route is simple, speed is low, and the lane is exclusively for the ship. In [2], also corresponding development in UK and in Finland is mentioned.

3. AUTONOMOUS OPERATIONS FOR SHIPS

The ultimate goal for autonomous operation for ships is to eliminate human crew and associated human failures committed. For a remote operated ship, the operation is assumed to be undertaken by a VTS. By eliminating the crew, it must be borne in mind that the crew in the VTS on shore is expected to be paid (mostly) by the ship operator. Hence, the cost saved is related only to the elimination of "safe manning", i.e. redundant logistic personnel. However, in this paper a safety comparison will only be made between a traditional ship, and a remote operated ship. Furthermore, it is assumed that the VTS control of each ship is undertaken in the same way as onboard a traditional ship. The VTS personnel is assumed to have the same competence level as on a traditional ship, as described in STCW.
The running of the ship with respect to engine operation is assumed to be taken over by automation.
Several problem areas are not discussed in this paper. Some of them are obvious;

- failure in data transmission to and from the ship
- failure of navigational equipment
- detection of traditional AToN
- restricted visibility
- leaks and ingress of water
- pilotage into ports, disturbing land light
- pirate attacks from sea or air
- 'severe weather' and 'other causes'

In particular, automatic pilotage into ports may be unacceptable because of potential damage to port facilities.

4. MODELLING OF POWERED GROUNDING

4.1 HUMAN FAILURE

A considerable fraction of causes for powered grounding human failure. In statistics in Norwegian waters from 1997 to 2014, the frequency of grounding in general (per sailed distance), as well as the human failure fraction is 0.6-0.7. The remaining fraction is caused by `technical failures', A disappointing result is that no reduction of grounding frequency has occurred.

4.2 MODELLING

A mathematical model for grounding is presented in detail in [1],based upon [4]. Practical applications of the model are given in [3] and [5].
Only an outline of the model is given here. The ship is on a `grounding course`. Without intervention, grounding will occur. On a traditional ship, intervention is to be made by the OOW, while on remote operated ships the intervention is executed by humans in the VTS on shore. It is assumed that the VTS control of each ship is undertaken in the same way as onboard a traditional ship, with a corresponding team, one VTS crew per ship. The instrumentation in the VTS `bridge` and on a traditional ship bridge are assumed to be identical. The means that relevant instrument signals must be transferred from the remote operated ship to the VTS, and intervention to avoid grounding must be undertaken by the VTS alone. AI is not involved.
As in [1], an event tree (ET) is used in the modelling. ET is a simple means to indicate the effects of various efforts to avoid grounding. As in [1], coarse and rather subjective numerical figures are assigned to the branch probabilities in the ET. The ET provides a tool that is simple to use, easy to change and which provide coarse numerical figures.
It is noted that the efforts are assessed to be only partly implemented. Hence, parts of the human failure fraction to avoid grounding will still remain.

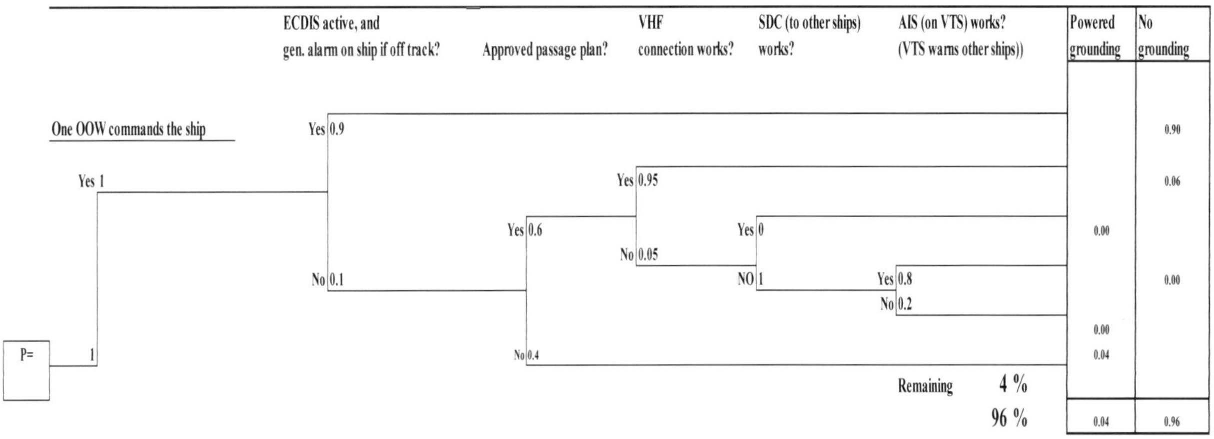

Figure 1. As today (2019), controlled by one OOW under normal conditions.

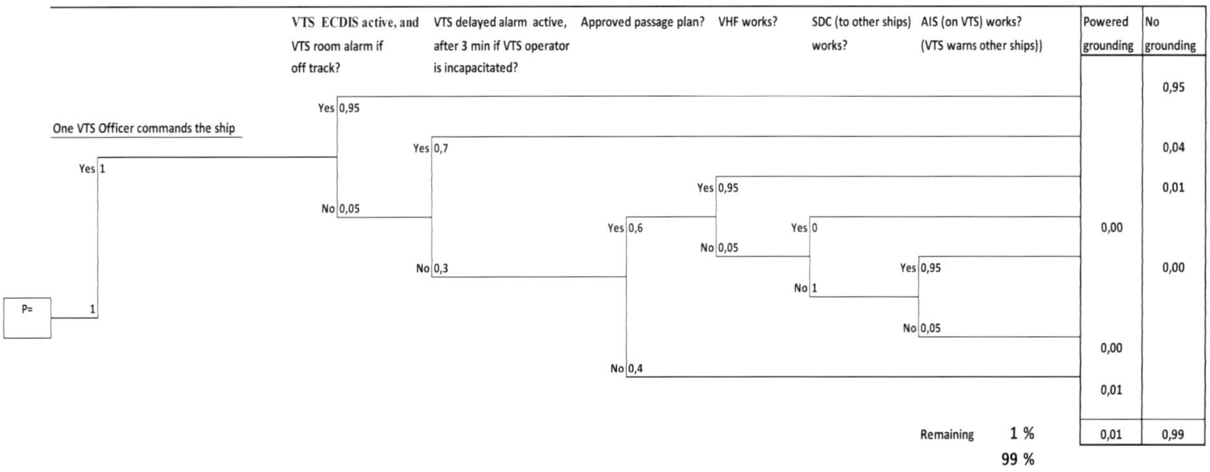

Figure 2. A remote controlled ship, controlled by VTS under normal conditions.

The results of the simple ET assessments are presented in Figure 1 and Figure 2.

The results are compared I Table 1.

Situation	Probability of grounding
From Figure 1. As today (2019), controlled by one OOW under normal conditions.	4 %
From Figure 2. A remotely controlled ship, controlled by VTS under normal conditions.	1 %

Table 1. Assessed reduction in human failure contribution for powered grounding for a traditional ship versus a remotely controlled ship

5. CONCLUSIONS

Numerous ship internal and ship external efforts to reduce the frequency of powered grounding caused by human failure have been implemented since the 1980-ties. From statistics, the result is very disappointing as the powered grounding frequency over the years has remained almost constant.

• In the coarse and simple ET assessment presented, the failure percentage for traditional ship control with one OOW is 4 %.

• a future significant reduction of human failure from 4 % to 1 % is indicated by leaving remote control to a VTS center ashore. This is on the condition that a strict TSS is implemented in coastal fairways. Without TSS it is difficult for VTS to define dangerous deviations from the course line.

To facilitate implementation of various types of autonomous ships, changes must be made to international laws, and to conventions like SOLAS and COLREGS. However, wide use of AI in coastal navigation is considered as not necessary and may cause increase the grounding frequency.

6. REFERENCES

1 DAHLE, E. AA., BENTZROD, K. LUND, N. H. `Ship internal and external efforts to reduce the human failure in ship operation` *RINA Conference on Human Factors, 28-29 September, London, UK*. 2016.

2. MARR, B.,` Intelligence Rather Than A Crew`. *Website Internet. Adobe Stock*. 2019.

3. The Coastal Administration of Norway. `The safety analysis of groundings and collisions in Norwegian waters 2014`. In Norwegian. *Report No. 1908Z31-6*. 2015.

4. KRISTIANSEN; S, `Safety Management and Risk Analysis`, *Elsevier* 2004.

5. DAHLE, E. AA., et al.. Risk Analysis of Gåsøyrenna (Oslo fjord). In Norwegian. *Safetec Report 2012-05-30*. 2012.

7. AUTHORS BIOGRAPHY

Dr. Ing., **E Aa Dahle, FRINA,** Marine and maritime consultant. He has been professor at the Technical University of Norway, and has i.a. worked with DNV GL. He holds a Class 3 DOC.

Capt. **K Bentzrod**, Master Mariner, presently at Hoegh LNG. He has been a master of ships for more than 30 years and has participated in development projects for the Coastal Administration of Norway.

Capt. **N H Lund**, Master Mariner and Maritime Pilot (Ret.). He holds a Master`s Certificate and has practiced as a Maritime Government Pilot in Norwegian waters for more than 35 years. He has participated in numerous VTS and other development projects for the Coastal Administration of Norway.

THE COMMUNITY OF SHIP-HANDLERS AND PILOTS: TRAINING, COLLABORATION AND GOOD PRACTICE TODAY AND IN THE FUTURE

R.J. Wild, Solent University, UK
M. Lützhöft, HVL, Norway

SUMMARY

Collective learning and knowledge sharing take place by a group of people, with similar interests, customs, language and repertoire, within a specific domain. Much of the community knowledge or memory bank is based on shared narratives and task norms developed over time and there is a constant thread of working and learning collaboratively, also known as a 'community of practice'. A survey was undertaken, designed to illuminate opinion and ascertain facts in a number of areas regarding community of practice theory. Ship-handlers and pilots were asked how they learned and acquired their skills, and the effectiveness of the various methods applicable to them. The survey yielded 113 responses with 61% fully completed. The participants highlighted the importance of a collaborative approach and indicated that over half of the necessary knowledge and skills is acquired whilst afloat, highlighting the importance of onboard training methods and tools to good practice.

ABBREVIATIONS AND ACRONYMS

CoP	Community of Practice
CPD	Continuous Professional Development
EMPA	European Maritime Pilots' Association
IMO	International Maritime Organisation
MASS	Maritime Autonomous Surface Ship
NI	Nautical Institute
OJT	On the Job Training
SOP	Standard Operating Procedure
STCW	Standards of Training Certification and Watchkeeping for seafarers
UKMPA	United Kingdom Maritime Pilots' Association
VCoP	Virtual Community of Practice

1. INTRODUCTION

The phrase 'communities of practice' (CoP) is an abstract concept coined by founding theorists Jean Lave and Etienne Wenger [1] after the analysis of research in the 1990s. The concept itself is as old as the history of man. There are many examples of CoP that can be professional (teaching, medicine), artisanal occupations (skilled craft workers) voluntary (civic groups) and exist in the virtual world [2, 3, 4, 5].

Lave and Wenger's research, which included a study of naval quartermasters, midwives and claims processors, identified three common features. Collective learning and knowledge sharing take place by a (i) group of people, with (ii) similar interests, customs, language and repertoire, within (iii) a specific domain. Much of the community knowledge or memory bank is based on shared narratives and task norms developed over time and there is a constant thread of working and learning collaboratively. Other researchers [6] build on earlier research and identified intermediaries as being a key component towards knowledge transfer and safety of operations within an organisation. These intermediaries are humans, artefacts and competing discourses. Competing discourses are documents concerning safety which provide a 'landing zone' for various components of an operation to be realised and co-joined. Components of an operation would typically be (i) purpose and function, (ii) the process norm and (iii) training and managerial oversight. The relationship between the intermediaries and the key role of practitioners (as opposed to technology or reference materials) in the transfer of knowledge, is illustrative of CoP theory and underpins the value of both elements to safety in dynamic and mission-critical environments.

2. PURPOSE

A survey was designed to collect subject matter expert views in a number of areas regarding ship-handling and pilotage community and to identify aspects of community of practice. Questions were based on the first author's experience as a pilot and ship-handler (over 22 years experience) and the survey was undertaken in two stages. The aim of the first stage was to make a preliminary assessment of the domain of pilots and ship-handlers and test questions for clarity.

Ship-handlers and pilots were asked how they learned and acquired their skills, and the effectiveness of the various methods applicable to them individually. The ship-handling CoP possesses unique skills that are acquired within the community and are not drawn from theory or reference material. Other skills may exist that emanate from outside the CoP and are acquired ashore for example and the survey was also intended to explore this. Participants were also asked what they understood by the phrase 'good practice'. The rationale for this is that whilst the good practice of today may or may not have relevance in the future, defining what it is and how it is transmitted is a necessary step in examining the value of the ship-handling CoP. In addition, questions were asked to gauge opinion regarding the future of the ship-handling and pilot community against the backdrop of the burgeoning autonomous era.

Many concepts in the survey were deliberately left undefined. For example, terms such as planning, good practice, mental models and mentoring are well known in the maritime industry and as the participants are all maritime professionals, it was decided to permit individual discretion.

3. METHOD

A survey was designed with the following aims:

- to gather data on ship-handling and pilotage training methods
- to examine the scope of ongoing ship-handling and pilotage training
- to compare the role of practitioners vs. management in ship-handling and pilotage training
- to review what good practice means to the ship-handling and pilotage community
- to analyse the causes of 'sub-optimal' events or manoeuvres and remedial action taken (if any)
- to explore how practitioners see their role in the future

4. INSTRUMENTS

This section introduces the survey questions and the options available to the participants.

Q1, Q2 and Q3. "Thinking about when you learnt 'what it is to be a pilot' and ship-handler, please select all the boxes from the list below that apply to your own circumstances. In the next question you will be asked to rank your choices i.e. indicate most effective to least effective".

Twenty-three options were provided (Table 1) plus the option 'other'. The options were derived from the first author's experience as a seafarer and pilot. A trial survey amongst a group of pilots was used to test the validity of the options and consider any others that emerged (see Section 5 Procedure). Q1 was not modified after the trial survey. Participants could select as many training methods as needed and were forewarned that Q2 would ask for all the selected options to be ranked in order of effectiveness. In Q3 participants who chose 'other' could enter free text.

Q4. "Thinking about how you would train a new pilot or ship-handler today, please use the slider to rank the effectiveness of traditional and new methods".

A scale was provided ranging from zero to 100, zero being in favour of new methods, and 100 in favour of traditional methods.

Q5 and Q6. "Thinking about when an act of ship-handling or pilotage did not go quite as planned or was sub-optimal (but was NOT due to a mechanical failure), please select all choices that apply. In Q6 you will be asked to rank the choices you made".

Six options were available, and participants could choose as many as they liked and then rank their choices in significance (Table 2).

Q7. "Please indicate from the choices below what ongoing training and continuous development as a ship-handler/pilot you have undertaken" (Table 3).

Q8. "Thinking about where improvements could be made to prevent sub-optimal ship-handling manoeuvres either as pilot or master, please make selections from the choices below".

The purpose of this was to compare Continuous Professional Development (CPD) and what participants believe is required to improve standards, following a sub-optimal event. The answer options are in Table 4.

Q9. "Consider the extent that ship-handling and pilot training is an 'establishment function' or primarily a collaboration between practitioners and learners".

A scale from zero to 100 was provided, zero indicating 'establishment function' and 100 'collaboration'.

Q10. "To what extent do you agree or disagree with the following statement: good practice within the community of ship-handlers and pilots is maintained by repeating what worked well on previous occasions within the community?".

A scale from zero to 100 was provided, zero indicating disagreement with the statement and a requirement to look outside the community, and 100 total agreement with the statement and that good practice is realised within the community.

Q11. "What does the phrase good practice mean to you?". This was a free text box.

Q12 and Q13. "In your opinion is autonomy a threat to ship-handlers and pilots in the future?".

In Q12 Participants were asked to use a Likert Scale using one of four choices; strongly agree, agree, disagree and strongly disagree. In Q13 participants were asked to describe in their own words why they made the selection.

Q14. Participants were asked to indicate their length of service experience by selecting one of four options.

Q15. Participants were asked to indicate their vessel type familiarity by selecting one of four options of various degrees of manoeuvrability.

Table 1: Training method options offered for Q1, 2, 3

Q1, Q2 and Q3. Training Method Options
Mental models imparted or shared by colleagues or other practitioners
Desk top models
Basic ship simulators (typically nautical college)
Sketches or diagrams
'Mentor' ship-master or other individual you learnt from
Trial and error
Reference books
Training videos or DVDs
Nautical college
Advanced ship simulator (not nautical college)
Published material (Nautical Institute for example)
Conferences/seminars
Manned models
Social conversations outside work
Informal conversations at work
Formal external incident reports
Internal incident reports
Observing other colleagues at work
Watching/looking at radar
Using ECDIS
Using PPU
Creating personal mental models and drawing on them later
Small boat experience (own boat or youth group etc)
Other

Table 2: Answer options for sub-optimal event causes

Q5. Sub-optimal pilotage or ship-handling manoeuvre choices
Planning
Incorrect information used in the plan
Unfamiliarity with task or manoeuvre
Environmental conditions (wind, current, tide) not as expected
Loss of awareness
Collaborative learning or mentoring has room for improvement
Other

Table 3: Answer options for training and development

Q7. Pilot and ship-handler ongoing training and development options
Reviewed my own ship-handling and pilotage acts
Reviewed ship-handling acts by colleagues
Read incident reports from flag state or own nation
Talked or corresponded with other ship-handlers/pilots from other organisations or companies
Read articles in relevant literature (reference books, magazines)
Read submissions about ship-handling on social media platforms
Attended dedicated ship-handling simulator training (not bridge team training)
Other

Table 4: Options for improving practice

Q8. Improving practice to prevent sub-optimal manoeuvres
Structured mentoring facilitated by your organisation
Better ship simulator training
Learning more theoretical knowledge as part of STCW for Certificates of Competency
More structured and realistic training as part of STCW for Certificates of Competency
More information about the berth, port of area
Other

5. PROCEDURE

The survey was initially distributed to a group of 26 pilots at a UK port and received a 71% return. Based on this, the survey was modified to improve question clarity and additional questions regarding good practice were asked. It was then sent to a wider audience and left open for a month. Contact was made with a number of bodies that represent professional mariners and marine pilots. The targeted groups were the Nautical Institute, the International Federation of Shipmasters' Associations, the UKMPA (direct email to members) and EMPA. Some basic demographics about experience and ship-type familiarity were gathered as part of the survey.

6. ANALYSIS

The survey yielded 113 responses with over 61% fully completed. The participants represented varying degrees of experience (Q14) and familiarity on a wide range of vessel types (Q15) and this is recorded in Table 5.

Table 5: Participant experience and vessel familiarity

Experience	**% of participants**
Less than 5 years	12%
Between 5 and 10 years	35%
More than 15 years	12%
More than 20 years	41%
Vessel type familiarity	**% of participants**
Conventional with no aids (no thrusters and standard rudder)	25%
Some aids (bow thruster, possibly non-standard rudder)	31%
Higher degree of manoeuvrability (multiple props, thrusters, possibly non-standard rudders)	23%
Extra manoeuvrability (azimuth thrusters, bow thrusters)	21%

The data from Q1 and Q2 concerning training method effectiveness were first categorised into nine themes: Mental models, mentoring, simulators & models, teaching/conferences, collaborating, electronic, doing the job, reference materials and 'Other'. The themes were determined by the researchers with the aim of reducing the number of options into clearly identifiable groups of similar tools and methods. For example, reference material includes training aids such as DVDs, published material and incident reports. Simulators at nautical colleges and independent facilities are counted together. Collaborating covers informal conversations both at and outside work plus observations with accompanying comments. Mentoring is a more formal arrangement.

The results are in Figure 1 and discussed in 7.1. Seven participants used the 'Other' option and the data were categorised into themes. The results were also grouped into training undertaken ashore or afloat, and regrouped again, into training undertaken collaboratively, self-managed, formally organised or blended.

Data regarding sub-optimal manoeuvres are presented in table 6, using percentage of participants selecting each one. The 'Other' option was used by 20 participants and qualitative data from the statements was analysed into themes. The results are in Figure 2.

Questions 3, 7, 8 used a scale between zero and 100 to indicate agreement with a statement or position. The data are reported as a number with accompanying text. Question 12 used a Likert scale of four steps, with end points of strongly agree and strongly disagree.

Questions 1, 4, 5 and 6 also offered free text options. The narratives were analysed qualitatively to reveal themes.

7. RESULTS AND DISCUSSION

The results show that more than half of the participants have over 15 years of experience. This level of experience places such participants at mid-career level and age approximately 40 years plus and means that the participants are experienced, and their answers should reflect that. More than 75% of participants are familiar with ships that offer a degree of manoeuvring aids (thrusters, non-standard rudders) and approximately one in five participants serve on highly manoeuvrable vessels which would lend themselves to being manoeuvred by the master or pilot, possibly without tugs.

7.1 Q1, 2, 3, 4 AND 7. THE SCOPE, NATURE AND EFFECTIVENESS OF TRAINING METHODS AND TOOLS

Questions 1, 2, 3, 4 and 7 are reported together as they concern training tools and methods. The 23 options were grouped in three ways, first into the 9 themes, secondly into training undertaken afloat and ashore and finally into collaborative, self-managed, organised or blended learning. Figure 1 shows the results of Q2 grouped in nine themes.

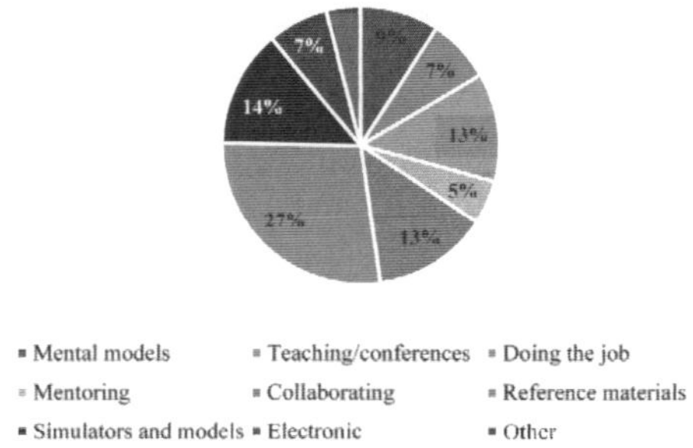

Figure 1: Themes of effectiveness of training methods and tools

After the first categorisation, the most selected method was reference materials (27%) which comprises books, other published material (such as NI output), incident reports and DVDs/videos.

The second categorisation indicated that more than 56% of the knowledge and skills is acquired whilst afloat (on the job, other vessels and manned models). Doing the job (15% OJT), mentoring and collaborating (15%) and using mental models (10%) are onboard activities that together account for 44% of all methods (afloat and ashore) and illustrate the importance of onboard training.

In Q4, traditional training methods such as observing practitioners, sailing on training vessels and other craft were selected by 37 of the 65 respondents when they considered how they would train a new recruit. Participants were allowed to determine what comprises traditional or new learning ship-methods or tools because they would have individual perspectives according to age and background. However, the majority of onboard methods would count as traditional whereas training ashore (simulators) and using videos and other reference material would be considered as newer methods (even though both have been in common usage at maritime training establishments since the 1970s).

In the third categorisation the 23 options were grouped into (i) those of a collaborative nature where practitioners learn together within a CoP, or mentors teach juniors, (ii) self-managed (e.g. studying reference materials), (iii) formally organised/ managed (simulators) or (iv) blended learning (e.g. maritime colleges which includes both). The collaborative group encompassed 77% of all learning. Question 7 queried the role of the establishment (zero) vs. practitioners (100) in training, and the average score was 62. Connecting these two responses indicates a strong role of collaborative learning.

Within the 'other' comments, a number of participants made a link between good practice and regulations and industry practice. Three examples are (i) *'good practice is in accordance with all regulations,* (ii) *good practice is obeying regulations,* and (iii) *good practice is following industry standards to the highest level'.* This suggests a degree of compliance with overarching governance. Such governance does exist for marine pilots in the form of IMO Resolution A960 Training of Pilots [7] and procedures or operational instructions issued by harbour and pilotage authorities. Whilst this governance is directed at pilots, there is no parallel body providing direction to other ship handlers. Because ship-handling is not exclusively undertaken by pilots, there is an argument to be made that similar governance could be extended to ship-masters who regularly undertake ship-handling, particularly within the confines of a port. Such governance would need to take into account the dynamics of ship operations which often requires a level of rapid decision making and even ad hoc creativity. There is often a separation between work as intended and actual work [8] and governance should not create a tension between the two.

Eight participants used the 'Other' option to describe personal training methods not listed within the options. All of the comments were related to onboard training (On the Job) on specific vessel types such as ro-ro ferries, tugs and bunker barges. This is another indication of the importance of training afloat. This is discussed below and reinforces the categorisation of the options into ashore and afloat themes in order to identify the value of each. One respondent (see 7.5) clearly identified the collaborative nature of pilotage and listed a number of stakeholders who play a part, possibly subconsciously, in advancing the skills of the port community as a whole.

7.2 Q5 AND 6. SUB-OPTIMAL MANOEUVRES AND REMEDIES FOR IMPROVEMENTS

Questions 5 and 6 are reported together as they concern sub-optimal manoeuvres and remedies for improvement.

Participants were asked to select and then rank the causes of sub-optimal ship-handling or pilotage manoeuvres on a scale of 1 (least likely) to 5 (most likely). The results of Q4 are in Table 6. 'Other' was chosen by 20 participants of participants and included a free text option. This option was used by 16 of the respondents and the narrative was analysed to look for themes. The results are in Figure 2.

Table 6: Causes of sub-optimal manoeuvres

Option	Selected by %
Planning	42%
Incorrect information	32%
Unfamiliarity with task or manoeuvre	23%
Environmental conditions not as expected	72%
Loss of awareness	33%
Collaborative learning or mentoring has room for improvement	13%
Other (free text option offered)	23%

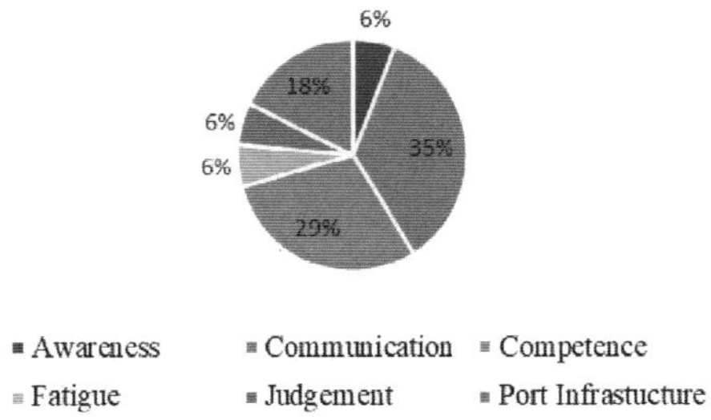

Figure 1. 'Other' causes of sub-optimal manoeuvres

35% of the 'other' causes are attributable to poor communications and 29% to competence. These are core elements of a community of practice and are closely aligned. 18% are due to port infrastructure and this includes the quality of resources and plant (tugs) and design factors (lock and berth fendering). These are outside of the control of the community itself, but familiarity with external shortcomings is to be expected. The three causes of fatigue, judgment and awareness (this was offered in Q4) all received one return.

7.3 Q7 AND 8. UNDERSTANDING THE SCOPE OF CONTINUATION TRAINING AND REFLECTIVE LEARNING

Figure 3 shows the ranked effectiveness of ongoing training and CPD for Q7.

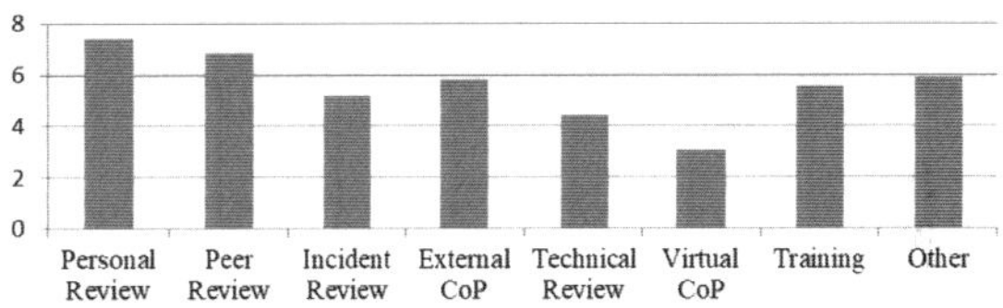

Figure 3: Ranked effectiveness of ongoing training and CPD

60% of the comments describing 'good practice' referenced various artefacts that together identify the community of practice and indicate its role. As reported in 7.1, external governance factors also play a role. Facilitating the spread of good practice, and the necessity of a collaborative approach and working with a number of stakeholders towards that objective, is identified.

In Q8 participants were asked for improvements that could be made following an incident (sub-optimal act) they had witnessed (or been involved in). The results are shown in Figure 4.

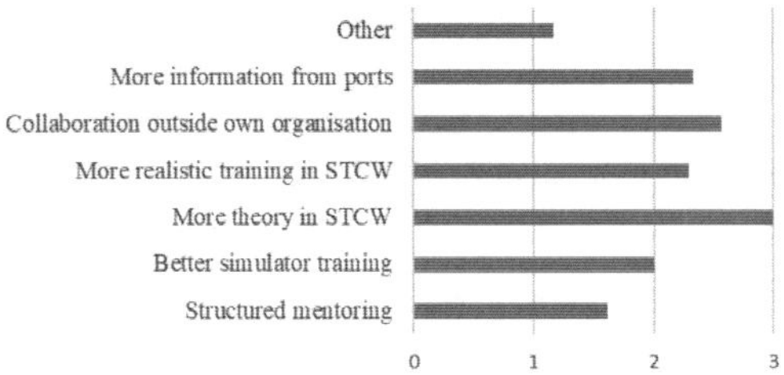

Figure 4: Ranked improvements for ship-handling and pilotage training.

The choice of 'Other' was used by 16 of the respondents. The results were analysed into six themes:

- Technology (access to live weather & environmental data)
- Resources (towage)
- Infrastructure (fendering)
- Competence (ship and shore)
- Over reliance on checklists
- Increase experience in unfamiliar manoeuvres and berths

Three 'Other' themes had an equal weighting: (i) crew competence, (ii) increasing training in unfamiliar manoeuvres and berths, and (iii) improving technology and resource at ports.

7.4 Q9. THE ROLE OF PRACTITIONERS AND MANAGEMENT IN TRAINING

This question was answered by 76 participants, 47 of whom indicated that pilot and ship-handling training is primarily a collaboration between practitioners and learners. This is an indication of the importance of collaborative learning and the role of the pilot and ship-handling CoP.

7.5 Q10 AND 11. UNDERSTANDING AND MAINTAINING GOOD PRACTICE

In Q10 participants were asked to indicate their agreement with the statement "good practice within the community of ship-handlers and pilots is maintained by repeating what worked well on previous occasions within the community". The average score was 54 (on a scale from 0 to 100) indicating agreement that good practice is maintained by the repetition of previous successful methods.

One respondent, who selected 'other' as a learning method and tool, stated that (pilots) should *'have more informal conversations with members of the port structure (tugs and linesmen for example) to build confidence but also to gain an understanding into how others work and numerous methods of working'*. Such conversations sit well with community of practice (CoPs) theory and the cascade of good practice. However, as CoPs are largely unregulated and not managed internally [9, 11], there is the possibility that bad practice can remain embedded.

In Q11 participants were asked to describe in their own words what 'good practice' means to them. The nature of 'good practice' was not defined, so participants were allowed to consider this from their own experience and contexts. The phrase was used deliberately instead of 'best practice' to clarify that 'best practice' implies that the process of learning, training and skill acquisition, plus reflection from any relevant incidents is complete and no further improvements can be made to 'the practice'. 'Good practice' leaves room for such improvements. The delineation was not challenged by any participants.

The word 'competence' featured in a large number of answers and a number of themes emerged which illuminate what competence is seen to be. These are shown in Figure 5. 'Competence' is frequently quoted as under-pinning good practice, and it is apparent that participants consider that it is closely aligned to non-technical skills. 'Independent thinking' is required but so is adherence to SOPs. This suggests a tension between competent practitioners and any external controlling factors, and this is in keeping with research findings [9, 10]. The management of CoPs is problematic, and this is a weakness because both good and bad habits can become embedded. 'Professional attitudes' are also required, but what they are and how they are acquired is not made clear.

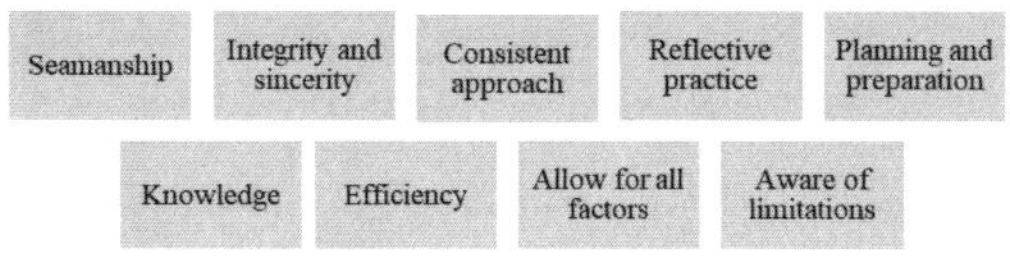

Figure 5: Community Artefacts

7.6 Q12 AND 13. THE ROLE OF SHIP-HANDLERS AND PILOTS IN THE FUTURE

Q12 was answered by 76 participants, 69 of whom also completed Q13.

57% of the respondents did not consider autonomy to be a threat. Inadequate technology (51% of comments) dealing with the local area (21%), and the cost (6%) were the most common reasons given. Other comments made included *'pushed by people with no knowledge' 'attempts by those without maritime knowledge to reduce pilotage' 'politics'* and *'the human element has to be part of ship-handling; this will never change'*. It is not possible to speculate on how familiar participants are with the latest developments in MASS technology due to access with projects which may be commercially sensitive. Nevertheless, 72% of comments are related to technology and how MASS would operate locally within ports and rivers.

Those who believed autonomy to be a threat said autonomy would reduce manning costs (38% of comments), technology could reduce human error (21%) and politics and lack of knowledge regarding the profession were drivers (10%). Other comments were made regarding vessel type (e.g. not passenger ships or dangerous cargo) and the location itself (some districts too complex or risks too high).

Q14 and Q15 gathered some basic demographic data which is reported at the beginning of this section.

7.7 SUMMARY

The results indicate that in spite of advances in technology, ship-handlers and pilots indicate that they have a role today and in the future. They also demonstrate that training is most effective when conducted afloat within a community of practice setting with trainees learning from practitioners. Reference materials and simulators also have a role to play. Ship-handlers and pilots also made a link between good practice and regulations and SOPs.

8. METHODS DISCUSSION

In Q1, participants were offered 23 options regarding training tools and methods, and then rank all that applied to them. Although it was made clear that Q2 would necessitate ranking the selected options, it is recognised that selecting large numbers of choices and then differentiating the effectiveness of each one could influence overall results. Only one comment was received in this regard. The average number of selections made was 8 and the mode was 6. The least was three replies with one method selected (a mentoring shipmaster (twice) and reference books). The largest number of selections was 21 and they had been ranked by the respondent.

A number of the training tools and methods options were associated with training afloat and it is possible that further nuances could be identified. This points to the necessity to capture live data during observations during the any further phases of this research.

9. CONCLUSIONS AND RECOMMENDATIONS

Whilst the study highlights the importance of onboard training methods and tools, training ashore, reference materials and particularly the use of simulators do have a role.

Collaborative approaches, along with associated mentoring is the most important method of training. This demonstrates that the community of practice is well established, (although participants may not call it this) which may reflect the considerable experience recorded by participants (more than 20 years by 41% of participants). This suggests that new practitioners should learn with as many members of the community as possible, which can be managed with a degree of control in a closed organisation such as a pilotage authority. However, in the wider context of the global ship-handling community, crew members in an organisation may live in several different countries, which makes establishing and maintaining a community harder. Shipping companies often train their officers on simulators but only so much can be achieved at training sessions lasting a few days.

There needs to be a determined effort to align all training with the community of practice ethos. This means accepting the preeminent status of the community afloat and ensuring that shore training (simulator and educational) formally access current practitioners to provide real life experience and skill.

A community of practice approach to cascading skills would seem to be an uncomfortable bedfellow with regulations and SOPs. Unwritten practice methods and customs handed down from one generation to the next is a dichotomy with the practice being controlled or orchestrated by formal procedures and regulations. However, as ‘practice’ implies a continuous learning process, with an element of trial and error, if mistakes are made that cause damage or injury then steps to prevent reoccurrence could include said regulations and procedures. The community itself must recognise the need for oversight and management which as mentioned can be opposed or rejected. This paper does not consider whether such ‘push-back’ from the community to authority is deliberate or embedded in sub-conscious levels, but nevertheless, proactive steps are required to promote ‘push-forward’ in the right direction.

The nub of the problem is that the profession of ship-handling and pilotage is undertaken by disparate groups and individuals all over the world. Representation of professional seafarers is varied and incomplete; there is no single mechanism to share training ‘tips’ and globally improve standards within the CoP. This means that a considerable ‘quantity’ of valuable insight, all of which contributes to good practice is not shared. This is to the detriment of the CoP; there should be a unified link between the formal and informal methods of training. Further illustration of this is the repeatably asked question as to whether ship-handling is an art or a science. This debate continues today. Certainly technology (such as GNSS and more powerful computers) enables more science to be used to describe ‘what ship-handling is’ but when local knowledge of a district, as espoused by 20% participants is considered, the *art* of ship-handling takes precedence.

The need for collaboration via multi-stakeholder forums and training opportunities is clearly identified, particularly within pilotage districts. In the future, these should be nurtured and the growth of virtual CoPs via social media should be encouraged. But there will be a need for knowledgeable moderation and leadership to ensure that there is an appropriate level of recognition and 'buy-in' from existing and potential practitioners. Social conversations outside work, ranked 13 out of 23 options are a good source of learning opportunities. Facilitating these is to be encouraged as this will nurture opportunities that might otherwise be missed. This could be problematic within a virtual CoP or VCoP. There are known issues with VCoP including knowledge hoarding (knowledge is power; do not share it), and culture (loss of face) [12]. Capturing, analysing and sharing insights made during social conversations about work practices could be valuable. This should be made clear to community practitioners, particularly in their formative years, although it is recognised that recordings can stifle honesty and spontaneity. Promoting awareness of how the community works and the associated advantages (and disadvantages) is fundamental to success.

10. ACKOWLEDGEMENTS

The authors would like to thank all the pilots and ship-handlers who completed this survey and contributed valuable insights.

11. REFERENCES

1. LAVE and WENGER, 'Situated Learning: Legitimate Peripheral Participation', *Cambridge University Press*, 1991.

2. NISTOR, BALTES and SCHUSTEK, 'Knowledge sharing and educational; technology acceptance in online academic communities of practice' *Campus Wide Information Systems*, 2012.

3. THRYSOE, HOUNSGAARD, DOHN and WAGNER, 'Newly qualified nurses – Experiences of interaction with members of a community of practice, *Nurse Education Today 32(2012) 551-555,* 2012.

4. VESCIO, ROSS and ADAMS, 'A review of research on the impact of professional learning communities on teaching practice and student learning' *Teaching and Teacher Education 24 (2008) 80-91*, 2008.

5. PATTON and PARKER, 'Teacher Education communities of practice: More than a culture of collaboration', *Teaching and Teaching Education 67(2017) 351-360*, 2017.

6. GHERARDI and NICOLINI, 'Learning the Trade: A Culture of Safety in Practice', *Organisation 7(2), 329-348,* 2002.

7. IMO and IMPA, 'Recommendations on Training and Certification and Operational Procedures for Maritime Pilots other than Deep-Sea Pilots, *IMO Res A.960*, 2004.

8. LOUKOPOULOS, DISMUKES and BARSHI, 'The Challenge of Routine Flight Operations', *Proceedings of the 12th International Symposium on aviation psychology,* 2003.

9. MITTENDORFF, GEIJSEL, HOEVE, DE LAAT and NIEUWENHUIS, 'Communities of Practice as stimulating forces for collective learning', *Journal of Workplace Learning 18(5), 298-312*, 2006.

10. HIDRETH and KIMBLE, 'Communities of Practice: Going one step too far?', *EconPapers HAL,* 2004.

11. PROBST and BORZILLO, 'Why communities of practice succeed and why they fail'. *European Management Journal 26(5), 335-347*. 2008.

12. ARDICHVILI, PAGE and WENTLING, 'Motivation and barriers to participation in virtual knowledge-sharing communities of practice, *Journal of Knowledge Management 7(1) 64-77*. 2003.

11. AUTHOR BIOGRAPHIES

Richard Wild is Senior Pilot at Harwich Haven Authority who provide pilotage services to five ports in the UK. He is responsible for piloting ships of 50 m LOA up to the largest container vessels in the world at 400 m LOA calling at the port of Felixstowe.

Margareta Lützhöft is a Professor in the MarSafe group at the Western Norway University of Applied Sciences. Her research interests include Human Factors, human-centered design and the effects of new technology.

THAT UNRELENTING 80 %

E. G. D. Verbeek Netherlands Pilots Cooperative, Netherlands

SUMMARY

80 % Human Error is an often used statistic in, among others, safety programs, publications, support for autonomous ships and even in the brochure for this conference.
By applying the concept "percentage Human Error" on various hypothetical scenarios this paper tries to discover the meaning, or lack thereof, of this indicator. It finds that mature systems are bound to have a high percentage Human Error and that automation will not reduce this percentage. It proposes to avoid use of the indicator "Human Error", and to decide on more meaningful indicators.

1. INTRODUCTION

Whenever I hear that 80 % of the accidents at sea are caused by Human Error; I am shocked.
I hope I offend no one, but I am shocked that after thousands of years of designing, building and equipping ships, the technology is apparently still so frail that one out of five accidents has technical causes.

In the beginning when I discussed this message, I was shying away from the logical conclusion. When someone would say: you cannot possibly mean that the percentage Human Error should even increase, I tried to agree: "no that was not quite what I meant". It took me a while to understand that actually I would be quite happy with an increase of percentage Human Error, as long as the number of accidents would decrease. I had to get used to that idea.

So in case you are a bit like me, I will tell the same message in a slightly different way: just suppose you want to buy a car, and the seller notices that safety is an important issue for you. Just suppose that he will say: this is not such a good car: almost 100 % of the accidents are because of Human Error. That other car is much safer: only 60 % of the accidents are because of Human Error, the other accidents are because of failing brakes, steering, wheels falling off, etc. Would you agree with this seller?
In fact what happens as soon as there is any technical problem? recalls!!!! In 2018, Toyota recalled more than 2,4 million Prius cars because of possible power stall in rare circumstances. Toyota declined to say if an actual accident did happen [1]....how many road accidents and how many due to a technical problem???? We're talking in excess of 99% Human Error.....and no one would want it differently. I am quite sure 20 % accidents due to technical causes would be absolutely unacceptable.

2. THE MEANING OF PERCENTAGE HUMAN ERROR

As a system develops, more and more is learned about the technical requirements: so less and less accidents will happen due to technical reasons. Mature systems are bound to have a high percentage Human Error. To illustrate in numbers: suppose that when a system is still in its infancy there are 1000 accidents: 500 technical and 500 Human Error, this gives 50 % accidents attributed to Human Error. As we learn what is needed to make sure nothing breaks, accidents due to technical reasons reduce to 50, and as we learn better procedures, accidents due to Human Error reduce to 250, this gives a great improvement in safety, a 70 % reduction in the number of accidents. However the percentage Human Error in accidents increases to over 83 % (figure 1). Although I don't have figures, I suspect that in the 17th century there was a much lower percentage Human Error: many accidents happened because of broken mast and rigging, rudders and the like.

.	Technical	HE	TTL	
initial	500	500	1000	50 % HE
mature	50	250	300	83,3 % HE

Figure 1

I think we can agree that the 80 % Human Error pops up everywhere. Googling "80 % Human Error" gave 1 billion 170 million results....I see the figure as support in presentations about Autonomous Ships (80 % Human Error, so *that* will be the reduction in accidents when ships are automated), in the brochure for this conference, in articles in all kind of magazines.

I have sometimes heard: "80 % Human Error, what are we going to do about it". Well, I have good news: if you would want to reduce it, that is very easy. Just build bad ships and put even worse equipment on board and you will see the percentage of Human Error in accidents reduce dramatically. The only downturn is that you will have to accept an increase in the number of accidents.

To illustrate my point, let's look at hypothetical ports Alfa and Bravo. In port A, 90 % of incidents are attributed to Human Error and in port B 60 %. Is port B safer then port A? Or are the 'humans' in port B better trained and skilled? Or does this indicate that port B has fewer incidents which are attributed to human error?
Well, it is possible that port A has a very good infrastructure, is relatively easy to enter, well organised and capable of forecasting winds and currents with good accuracy. Incidents only happen with an average frequency of 2 per 1000 movements, so incidents attributed to Human Error occur at a rate of 1.8 per 1000 movements.

And port B may be an atrocious port with unpredictable wind shears, changes of current and eddies. Incidents happen at a rate of 6 per 1000 and despite having a lower *percentage* of incidents caused by Human Error, in terms of *numbers,* incidents due to Human Error occur twice as often as in port A, at a rate of 3.6 accidents per 1000 movements (figure 2).

	Port A	Port B
% incidents attributed to HE	90 %	60 %
Incidents per 1000 movements	2	6
Incidents due to HE per 1000 movements	1,8	3,6

Figure 2

On the other hand, may be ships calling at Port A are well maintained, however the pilots lack some training. And in port B there might be well trained pilots, but the ships that are calling are in a bad state, which causes technical break downs (figure 3).

	Port A	Port A'	Port B	Port B'
% incidents attributed to HE	90 %	90 %	60 %	60 %
Incidents per 1000 movements	2	4	6	3
Incidents due to HE per 1000 movements	1,8	3,6	3,6	1,8

Figure 3

So the fact that 90 % of accidents in port A are attributed to Human Error and in port B 60 %, does not say anything about the overall safety of the ports, nor about which are the issues in the ports. It even does not say anything about the frequency with which incidents attributed to Human Error occur. I'm afraid we have to say that it does not say anything at all.

Even that reputedly the percentage Human Error stubbornly stays at 80 % does not say much: according to the figures released by Allianz the number of total losses of ships over 100 GT greatly reduced from 207 in 2000 to 46 in 2018, while in the meantime the number of ships has increased [2], So if we accept that in 80 % of the cases these losses where attributed to Human Error, the number of losses due to Human Error reduced at the same rate….so safety improved while there was a constant percentage Human Error…..

3. ARE HUMANS THE WEAK LINK?

But perhaps the fact that 80 % of accidents are attributed to Human Error does prove something. Many say that it proves that humans are the weak link.
This could lead to the conclusion that we must try to automate the human out of the system to make the system safer.

Now that seems a bit of a one-sided conclusion to me. We meticulously document and count the number of times that people mess up. We even call it 'Human Error' when the automation couldn't cope and people did not take the right corrective action. But we do not document the cases in which people repaired or recovered problems with equipment. We do not count the number of times that people successfully took over when automation could not cope. Yet to know if people are an asset or a liability, it is necessary to compare these two sides of the equation.

For arguments sake: let's take autonomous, unmanned ships, I hope you don't think that I am too negative by assuming that, once in a while, something will still go wrong. Here, if put at the same level, on board of the ships, 100 % of accidents must be attributed to technical/automation causes. If 80 % Human Error proves that humans are the weak link, and should be automated away, with 100 % technical/automation error, automation must equally be seen as the weak link, and should be manned away.
This is clearly nonsense. Automation has greatly contributed to higher safety levels. I hope that we can all agree that it is similarly non-sense to say, based on the 80 % Human Error, that humans are the weak link and should be automated out of the system. The only way to know if one way is more safe then the other is to have comparative data on how many accidents happen per meaningful unit.
The fact that in the first case *humans* are considered to be responsible for the great majority of accidents and in the second case *automation*, only shows that in the first case humans were in control and in the second case machines, no more, no less.

But suppose we would try to reduce the percentage Human Error by automation with human supervision, would that work? Well, actually automation with human supervision will automatically lead to a higher percentage Human Error. An example: mooring a cruise vessel with a joy stick. Just before mooring a gust hits the vessel. If the vessel is successfully moored, the captain has done well, whatever he did. If there is damage, and the captain took over on hand, the company will say: old fashion captain, you should put more trust in the equipment we provided (counts as Human Error) and if the captain would have kept it on joy stick: captain, we pay you a fortune exactly for these circumstances: you should have taken over (counts as Human Error). So when things go sour: it is always counted as Human Error…no reduction of percentage Human Error…
The difficulties humans have with monitoring, or skill loss associated with automation, I'll leave for another time.

I have seen a DVD aimed at making crews more aware of the need to manage pilot error [3]. The analysis of the case studies starts with the state of the vessels: some may be older, but always mechanically sound and with navigational equipment in good working order. Although

not mentioned, I do not doubt for a moment that the ships and their equipment are fully certified. Then the DVD comes to the conclusion that it is obvious that there can be no technical reasons for the accident and it searches for what the operators have done wrong.
Of course it is also possible to look at things from a different angle: the crew and pilot were well trained and experienced, fully certified, so what was wrong with the ship and/or equipment?

One case deals with the grounding of a large container vessel. The pilot used the radar with an off-centre display to maximise the view ahead. When making a large alteration of course in fog, he misinterpreted the Electronic Bearing Line by reading the azimuth ring around the picture, instead of the electronic numerical readout. Because the radar was off-centre and the azimuth ring was centred, there was a discrepancy between the two and reading the azimuth ring gave the wrong information.
Based on the first perspective, the analysis found that this was a clear case of 'Human Error'. The pilot should have known better. And the crew should have watched him more closely.
Based on the second perspective one might come to the conclusion that the equipment invited the error. It would be very easy to remedy this by either adjusting the azimuth ring to conform with off-centre display, as one radar manufacturer does, or just switch the azimuth ring off as soon as off-centre is selected (technically relatively easy to do).

At this point I ought perhaps to define 'Human Error' a bit more sharply. "Human Error" is generally used to mean 'operator error'. If we also used it for lawmakers, designers, builders etc it would be obvious that 100 % of the accidents are caused by Human Error.
But can it always be labelled 'error'? In complicated systems like shipping, there are multiple goals, such as efficiency and safety. To fulfil them there are at times conflicting demands. People have to balance these conflicting demands and have to make compromises [4]. If one could call a system that complies 100 % with all goals "perfect", then these compromises lead to imperfections in legislation, design, manufacture, guidelines, procedures and all other aspects involved in operating a ship. The operator inherits all these imperfections and has to cope with them. He also has to make his own trade-offs to meet the multiple goals at his level. Generally that works well: operators usually succeed in developing strategies that make the system work in, let's say 99.5 % of the time, despite all these conflicting goals and inbuilt imperfections. When it goes wrong there is a fair chance that the operator gets blamed for doing exactly what he has been doing all the time to make the system work smoothly. Only this time it went wrong because the circumstances were different. Putting the blame on the operator does not necessarily improve safety because there is a fair chance that a next operator will make a similar mistake.

Using both perspectives would give the best result, looking at improving procedures /teamwork and improving the equipment so that it is designed to be used by humans, something which is not always recognised by manufactures.
To quote a chapter heading in a book by prof. Dekker: 'It is not Individuals *or* Systems, but Individuals *in* Systems' [5]. So I am very happy with the drive towards Human Centred Design.

4. CONCLUSIONS

May be we should strive to increase the percentage of accidents 'caused by Human Error'. I would say: let's try to get it up to at least 95 %. That is not something terrible, but something good, if it means that the number of accidents goes down. It is a sure sign that we have given the people at the sharp end the right equipment.

But actually I propose to stop using 'percentage of accidents caused by Human Error' as indication of safety or as information to guide development. As I have shown, against popular belief, an increase in percentage Human Error in accidents could in fact indicate an increase in safety, but not always. And a decrease in percentage Human Error suggests a decrease in safety, but not necessarily so. Even if the percentage stays the same, that does not give any indication of how the safety situation develops.
And as we hope that in future 95 % of the accidents will be 'caused by Human Error' and generally agree that currently 80 % have 'Human Error' as a 'cause', stating that an accident is caused by Human Error adds very little useful information. So let's stop using this meaningless and unhelpful, even potentially misleading yardstick. Instead let's try to develop meaningful terms that are helpful to measure changes in levels of safety. In the air industry a much used yardstick is incidents per million departures [6]. Little happens mid-flight, so if you measure per hour as was done in the past, safety would increase without any change as flights become longer.

For instance: let's measure accidents in ports per 10.000 movement. Let's measure accidents at high seas per 1.000.000 miles. Even if the figures are not known with a great degree of accuracy, they would still give a far better indication.
And please don't replace Human Error, Human Factor etc by terms as 'loss of situational awareness' which no doubt occurs for 95+ % of the groundings and collisions. If the person had adequate situational awareness, most likely the accident would not happen...
Instead, let's try to understand and describe the interaction between operator and machine and between man and system, and the reasons for breakdown's in these.

Let's try to understand why the operator has acted the way he did. In virtually all cases, the operator did not plan to mess up. He wanted to do a good job. What he did 'made sense to him given the goals, knowledge and attentional demands at the time' [6].

If we don't understand why he did the way he did, we have the wrong perspective and have to change viewpoint.

My proposals

- Aim for greater safety, accept that percentage Human Error might even increase, and the term has little added value, so avoid using it
- Develop other, more meaningful indicators
- Stop separating Human Error and technical causes, and look at the total picture

5. REFERENCES

1] USA TODAY Published 11:18 AM EDT Oct 5, 2018, retrieved 23 January 2020 from https://eu.usatoday.com/story/money/cars/2018/10/05/toyota-prius-recall/1532120002/

[2] Allianz, The Safety and Shipping Review 2019; retrieved 23 January 2020 from https://www.agcs.allianz.com/news-and-insights/expert-risk-articles/shipping-safety-worst-accident-locations.html

[3] DVD American Club, *Stranger on the Bridge*. 2007.

[4] Woods, D et al (2010) *Behind Human Error (2nd edition)*. Farnham, U.K.: Ashgate

[5] Dekker, S (2007). *Just Culture*. Aldershot, U.K.: Ashgate.

[6] Dekker, S (2006). *the Field Guide to Understand Human Error*. Aldershot, U.K.: Ashgate.

[6] Kaspers, S et al (2019), *How Does Aviation Industry Measure Safety Performance?*
Retrieved dd 27 January 2020 from https://www.researchgate.net/publication/331724708_How_does_aviation_industry_measure_safety_performance_Current_practice_and_limitations

6. AUTHORS BIOGRAPHY

Ed Verbeek

Ed Verbeek has been educated as Master Mariner and has a Master's degree in Human Factors and System Safety issued by Lund University. He has been elected Fellow of the Nautical Institute.

Ed has sailed as mate for about 12 years before becoming pilot, continued with occasional trips as master while a pilot and after retiring as pilot still does some relief trips as chief mate or master.

He has been Amsterdam pilot for 26 years, and has been Vice President and Training Coordinator for this region.

He has been involved in different forms of Bridge Resource Management since 1993 and has been instructor shiphandling for CSMART, the training institute of Carnival Corp.

Ed is active as consultant, among others working as Advisor Research & Development for the Netherlands Pilots Cooperative, was assessor at promotion assessments (captains and deputy captains) for Cunard, P&O and Princess, and consultant for port studies at the CSMART simulation centre.

PERCEPTIONS OF THE WEST AND THE EAST AND NATIONAL CULTURE ON THE HUMAN FACTORS

Y. Kwon, Chosun University, KOREA

SUMMARY

Researches and discussions on the human factors are worth for better safety of, in particular, transportation industries. In fact, it may deserve more attention as the modern industries look for AI or autonomous operation in the near future. Meanwhile, it has also been anticipated that such researches and discussions are rather one way along the technical and engineering approaches. Safety, being also an issue of cultural scrutiny, this paper tries to show the importance of perception of the thoughts between the east and the west, then the role of national culture referring relevant researches appeared recently.

1. INTRODUCTION

The author had opportunities at times working in group meetings for the safety of ships, but ended up finding huge gaps in the perception of the issue in his home country, Korea, and then also between the west and the east [1]. He continued to find the need to consider Hofstede's national culture [2][3]. Simple conclusion out of the research is that the approaches and measures found and recommended in a specific region/country may not work efficiently in another region/country due to differences in culture, that is, values and practices.

There have been researchers who paid attention on the cultural issues in the safety analysis, such as Kaplan [4], Hendrikse et al. [5] and so on during the last 20 years. Recently, Yoder and Messer-Bookman, noticing seafarers from diverse cultures work and live in contained environments, expressed that culture, such as gender, power distance, sensory perception and cognition etc., underlies the entire error chain of events [6]. They suggest the importance of developing deeper understandings of how culture influences the way in which we experience and interpret our world in order to improve our ability to manage challenging (safety) circumstances at sea.

1.1 SCOPE OF THE RESEARCH

This paper is to look into the way of perception and mechanism of behavioral manner, consulting results of the previous serial works of the author, that is, on (1) western ontology and eastern relationism [1] and (2) applications of Hofstede's national culture [3][7]. The results and discussions are rather simple as it comprises comparison between selective countries of high and low accidents rates along the cultural dimensions. Then fundamental social scientific terms relevant to safety issues on each cultural dimension are further examined to see the effects and cause. In the context, the author has to admit not only the personal limitation of scope and development, but also it is such a complexity of social science and socio-technology. Nevertheless, analysis of the implication of the key words in each national cultural dimension, considering together with the influence of essential thoughts of the west and the east, would be of use for the basic frame and structure of the safety policies of a country in need. It is anticipated that further long-term scrutiny and researches are necessary by social scientists together with safety experts.

1.2 CRITIQUE ON THE HOFSTEDE'S NATIONAL CULTURE MODEL

There are critiques on the Hofstede's model of the national culture (see e.g. [8][9][10]). For instance, McSweeney asserts Hofstede's claims are excessive and unbalanced [9]. Well, the author may not be at the exact position to deal with the discussion in detail, but it is surmised that the user of the model may have to be cautious, as the spectrum of the analysis and description is wide enough. Nevertheless, it is a good opportunity to look into the model if the effect of individuals/professionals and organizational culture may not be of any practical help for better safety in certain countries. Furthermore when either national or international level of investigation is required, Hofstede's model would be a good starter to begin. In fact, the issue may be worth incorporating with 'Just Culture' [11].

2. THE WEST AND THE EAST AND NATIONAL CULTURE WITH SAFETY

2.1 THE WEST AND THE EAST

The western ontology considers that individual being is the fundamental unit and therefore is given to the individual existence. It maintains a motion principle of constantly strengthening itself. Whereas, every being on earth exists within a network of relations in the eastern perception [1][12]. Kwon discussed on the issue and proposed combining merits from both sides, overcoming ensuing difficulties [1]. It is also suspected that such conceptions of the west and the east are similar in practice with individualism and collectivism (IDV) as well as short-term and long-term orientation (LTO), respectively, in Hofstede's model.

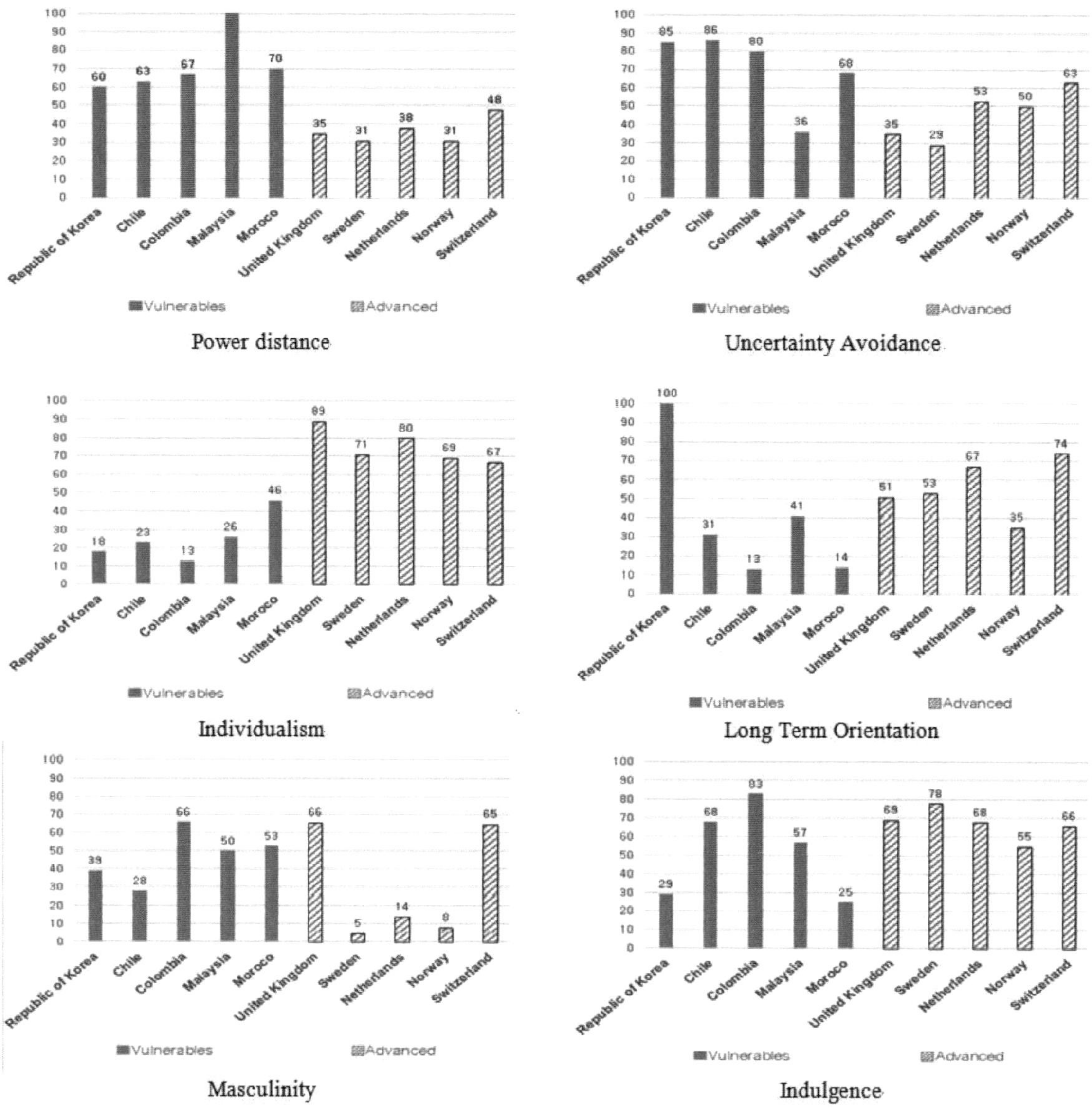

Fig. 1 Comparison of national cultural dimensions in case of road traffic accidents [6]

2.2 NATIONAL CULTURE WITH SAFETY

A simple application of the Hofstede's model as to road traffic accidents was carried out to show the effects of national cultural dimensions [3]. From the results shown again in Fig.1, it was found that among the 6 dimensions, (1) power distance, (2) individualism-collectivism and (3) uncertainty avoidance appeared to be related with road traffic accidents. Then with the help of key differences between small and large dimensional indices, key words relevant to safety behaviour, such as communication, tolerance to others, are drawn to further scrutiny for measures and practical policies. Note that Hofstede also states that stronger uncertainty avoidance means faster driving [2]. Later on, additional data are prepared to see the same tendency in marine accidents and shown in Fig. 2 [7].

2.3 RESULTS

Selections of key words relevant to safety in each dimension are as follows;

(1) Power distance: (in)equality, status, authoritarian value, hierarchy, (de)centralization, dialogue, violence

(2) Individualism-Collectivism: I (we), value standards, guilt (shame) and loss of self-respect (face), in(ter)dependent, speak up, purpose of education, diplomas, management of individuals (groups), , task (relationship) prevails over relationship (task), privacy, laws

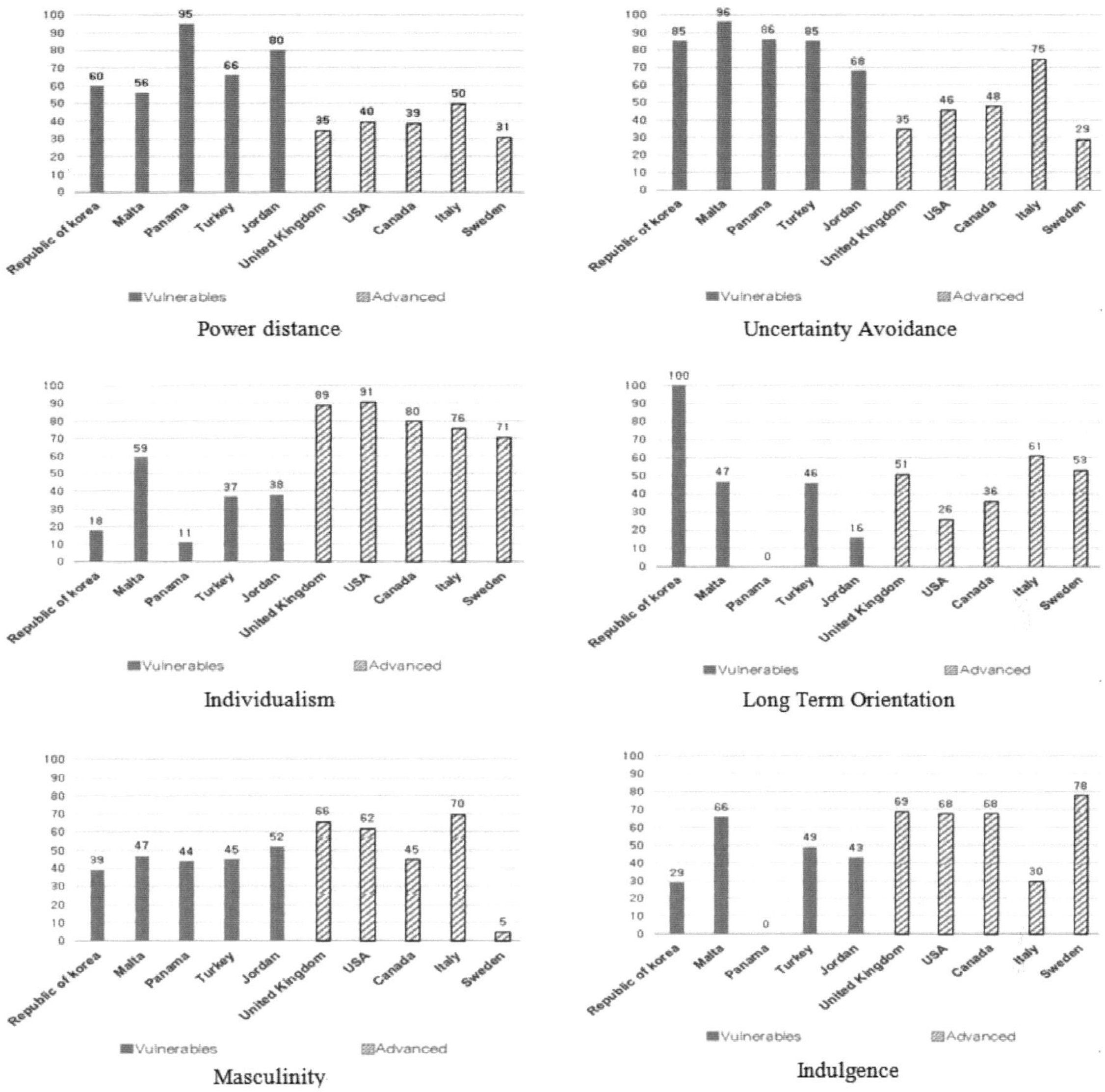

Fig. 2 Characteristics of national culture dimensions in case of marine accidents [9]

and rights, human rights, ideologies of individual freedom (equality) prevail over ideologies of equality (individual freedom)

(3) Uncertainty avoidance: uncertainty is a normal feature (continuous threat) and accepted (must be fought), low (high) stress and anxiety, aggressions and emotions, agreeableness (neuroticism), weak (strong) superego developed, what is different is curious (dangerous), relaxed (stressful) family life, fewer (more) people feel unhappy, students are comfortable with open-ended (in structured) learning situations and concerned with good discussions (the right answers), results are attributed to a person's ability (circumstances or luck), reading books, (no) more rules than strictly necessary, tolerance for ambiguity and chaos (need for precision and formalization), belief in generalists and common sense (in experts and technical solutions), focus on decision process (decision content), fast (slow) results in case of appeal to justice, citizens are (in)competent toward authorities, citizen protest is accepted (should be repressed), citizens trust (are negative toward) politicians, civil servants, and the legal system, high (low) participation in voluntary associations and movements, the burden of proof for identifying a citizen is on the authorities (citizens should be able to identify themselves), less (more) corruption, positive (negative) attitudes toward young people, tolerance even of extreme ideas (extremism and repression of extremism), lower (high) risk of violent intergroup conflict, a

tendency toward relativism and empiricism (grand theories) in philosophy and science, scientific opponents can (not) be personal friends

3. DISCUSSION – SAFETY CULTURE AND MARITIME SAFETY

3.1 POWER DISTANCE

Power distance (PD) is well known factor which impacts people's attitudes and behaviours, affecting every interaction [6]. The key words themselves show up the issue - it is matters of dialogue and misuse of power [2][3]. As to the dialogue, it is known that eastern peoples are good at reading others minds. Well it has got merit at times, but risky and dangerous, before proper dialogue. As Hofstede speculated, the use of power should be legitimate, which is based on formal position and expertise, not tradition or family. He further examines that Korea is a slight hierarchical society, where people accept a hierarchical order. In irrational hierarchy society, the need of self-actualization may lead oneself to distorted status. It is noted that in small power distance community, (1) hierarchy means an inequality of roles, established for convenience, not for status nor power, (2) status symbols are suspect and (3) organizations are supposed to have structured ways of dealing with employee complaints about alleged power abuse. The present reformation movement of the prosecutory affairs in Korea may hopefully redeem the political muddling up as well as people's daily life. Misuse of power evidently appears in traffic activities [13].

3.2 INDIVIDUALISM-COLLECTIVISM

In a collectivist culture, loyalty is paramount and overrides most other societal rules and regulations, where offence leads to shame and loss of face instead of guilt and self-respect respectively, which recurs violation in safety practices - the most contributing factor in traffic accidents [2][14]. It is worth noting that shame is social in nature, whereas guilt is individual; whether shame is felt depends on whether the infringement has become known by others [2]. Pursuing face may lead to showing off and, then, again to oppression [15]. Nusbaum claims that, although there are merit, the feeling of shame on and disgust at someone comes from (s)he can control others and impairs responsibility of others, respectively [16]. She further states that pressure on common good may infringe on personal freedom and diversity. By the way, to the author's experience, a Westerner uses the term 'shame' often in daily conversations, and the message seems to be rather different in practice. In the collectivist family, children learn to take their bearings from others when it comes to opinions [2]. This goes further in class in school. Children, students and even pressmen seldom raise hand to questions, which is also due to the purpose of education, not how to learn but how to do. It is irony that collectivist cultures encourage an interdependent self, but have little regard for others.

3.3 UNCERTAINTY AVOIDANCE

Uncertainty avoidance (UA) is considered as an intriguing and important dimension. Extreme ambiguity creates intolerable anxiety. UA being defined as the extent to which the members of a culture feel threatened by ambiguous or unknown situations, it is strongly related with the degree of anxiety [2]. One typical example is high suicide rates outcome of anxiety in a society, then another is, paradoxically, faster driving, which means more fatal accidents, thus more risk. However, it is a familiar risk. It comes with sense of stress of urgency. Stronger UA countries show a priority of saving time over saving lives [2], which is the case of Korea. Making it worse, the fact that pedestrians count for 39.9% (OECD mean value is 19.7%) of all fatalities, whereas drivers 19.2% (OECD mean value is 46.6%) [17] is dreadful and must be a great concern, let alone the total fatality. In recent years, Korean government is trying hard to boost safer environments, and lower speed limit in city centres has been introduced. In short, the anxiety component of UA leads to noticeable differences between strong and weak UA societies. We may refer to De Botton, who states that (status) anxiety is due to (1) lovelessness, (2) expectation, (3) meritocracy, (4) snobbery and (5) dependence and (1) philosophy, (2) art, (3) politics, (4) religion and (5) bohemianism would be solutions [18].

The implications of UA for the relationship between authorities and citizens differ from those of power distance. In high PD countries authorities have more unchecked power status, and material rewards than in low PD countries. In high UA countries authorities are deemed to have more expertise than in low UA countries. The inequality in this latter case is, we may note that, not in the power but in the competence of authorities versus other citizens [2].

The fundamental difference between higher and lower UA would be the sentiment of 'dangerous' and 'curious', respectively, when encounters 'difference'. It may also be noted that lower self-ratings on health in UA cultures are reflected in higher self-ratings on unhappiness. The emotional need for laws and regulations in a strong UA society can lead to rules or rule-oriented behaviours that are purely ritual, inconsistent, or even dysfunctional [2].

3.4 SUMMARY

Before summarizing discussions, it may be noted that, as Hofstede et al. describe, studying differences in culture among groups and societies presupposes a neutral vantage point, a position of cultural relativism. They rightly refer Claude Levi-Strauss as follows [2];

Cultural relativism affirms that one culture has no absolute criteria for judging the activities of another culture as "low" or "noble". However, every culture can and should apply such judgement to its own activities, because its members are actors as well as observers.

In brief, national safety policies for safer society or country require scrutiny of cultural analysis of its own society/country. It may include selecting and dealing with social impact factors and their causes and effects to establish social calibrations. For instance, to name a few, (1) enhancing transparency in using the power and administration, which makes rooms for trust [19], (2) reducing anxiety and violence, which in turn increases tolerance, (3) reaffirming the purpose of education, (4) opening up the dialogue, which will also provide situation awareness, empathy and care, should be in line with foremost priorities in the long run. We may also note that all 3 cultural dimensions of PD, INV and UA are related with violence and physical/verbal abuse which critically attribute to against the safety practices.
As an extension of the discussion, such a policy may also be effective to be discussed and tried at IMO for safer shipping, crew and marine environment along with the agenda of Just culture for the various member countries [11].

4. CONCLUSIONS

(1) Both of the west and the east have been evolving to develop grammars of their own ways of living for individuals/professionals, organisations and nations [1], thus the cultures respectively, but inevitably for the same goals and achievements.
(2) Study of cultural differences can help its own community/country as well as international harmony through understanding itself and one another.
(3) Hofstede's model is of convenient use to figure out the problematic issues which become causes and effects on (safety) attitudes and behaviours with the help of cultural anthropology, social science and human nature.
(4) With philosophy, culture would be getting more attention as the progressive technology shows almost no limit in the near future.

5. ACKNOWLEDGEMENTS

The author would like to acknowledge some ideas and assistance provided by senior students Mr. Tae Soo Jung and Mr. Dae Hee Seong. He also appreciate Dr, Professor Carol Yoder at Trinity University in San Antonio, Captain, Professor Emerita Tuuli Messer-Bookman at California Maritime Academy and Professor Sanja Bauk at Durban University of Technology for their constructive communication and encouragement.

6. REFERENCES

1. Kwon, Y. 2012. A Review of the Safety Culture for Developing Maritime Safety. Proceedings of the Conference on Education & Professional Development, Southampton, UK
2. Hofstede, G. Hofstede, G.J. & Minkov, M. 2010. Culture and Organizations: Software of the Mind. McGraw Hill
3. Kwon, Y. 2019. A Review on the National Culture and its Effects on the Transportation Safety Perspectives. TransNav: Int. J. on Maritime Navigation and Safety of Sea Transportation, Vol. 13 No. 2, June
4. Kaplan, M. 1995. The culture at work: cultural ergonomics. Ergonomics, Vol. 38, No. 3, pp. 606-615
5. Hendrikse, E. J. McKinney, A. E. McCafferty, D. B. and Poblete, B. R. 2000. Human Factors Engineering and Cultural Calibration for an Offshore Platform design: are there lessons for the maritime industry? Human Factors in Ship Design and Operation, Sept 27-29, London
6. Yoder, C. and Captain Messer-Bookman, T. 2019. Same facts, different perceptions, Seaways: The Int. J. of the Nautical Institute, September
7. Kwon, Y., Jung, T. S. and Seong, D. H. 2019. Perspective of National Culture for the Development of Just Culture as to Transportation Safety. Autumn meeting, The Korean Society of Marine Environment & Safety, Nov. 28 & 29 (in Korean)
8. https://en.wikipedia.org/wiki/Geert_Hofstede
9. McSweeney, B. 2002. Hofstede's Model of National Cultural Differences and their Consequences: a triumph of faith – a failure of analysis. Human Relations, Vol.55, No.1, pp89-118
10. Cale, M. H. 2011. Cultural Differences and Traffic Safety: Useless knowledge or the key to the solution? CogniTo Ltd.
11. IMO MSC 88/16/1. 2010. 'Role of Human Element, Just Culture – Essential for Safety'. Submitted by the UK
12. Shin, Y. B. 2004. Lecture: My reading of oriental classics. Dolbegae (in Korean)
13. Kang, J. 2011. Korea, an Extraordinary Country. Inmulgwasasangsa (in Korean)
14. Lee, S. 2019. A Study on the Estimating Licence Demerit Point about the Items Causing traffic Accidents. J. Korean Soc. Transportation, Vol. 37 No. 5, October
15. Park, H. and Park, J. 2019. Get old reading all the way. Sideways (in Korean)
16. Nusbaum, M. C. 2015. Hiding from Humanity: Disgust, Shame, and Law. Minumsa (in Korean)
17. Jung, J. 2019. Future Works for Tolerant Safe Society. Joint Conference of Safety Policy of Disaster: the Road to Tolerant Safe Society, Korea Institute of Public Administration, Statistical Research Institute and UNIST, December
18. De Botton, A. 2011. Status Anxiety. Eunhaengnamu (in Korean)
19. DeSteno, D. 2018. The Truth about Trust. Woongjin ThinkBig Co. (in Korean)